OXFORD LATIN COURSE

TEACHER'S BOOK

MAURICE BALME AND JAMES MORWOOD

Oxford University Press 1987

Oxford University Press, Walton Street, Oxford OX2 6DP
Oxford New York Toronto
Delhi Bombay Calcutta Madras Karachi
Petaling Jaya Singapore Hong Kong Tokyo
Nairobi Dar es Salaam Cape Town
Melbourne Auckland

and associated companies in
Beirut Berlin Ibadan Nicosia

Oxford is a trade mark of Oxford University Press

ISBN 0 19 912093 5

Acknowledgements

The authors would like to express their warmest thanks to George Littlejohn of Smithycroft Secondary School, to Professor E. J. Kenney of Peterhouse, Cambridge, and to Dr Jonathan Powell of Newcastle University for their generous and helpful advice.

The authors and publishers are grateful to the following for permission to reproduce extracts from copyright works:

The Aeneid of Virgil, translated by C. Day Lewis: The executors of the estate of C. Day Lewis and The Hogarth Press; *Horace Odes II, 7*, translated by James Michie: James Michie; *The Memoirs of Hector Berlioz*, translated and edited by David Cairns: David Cairns.

Phototypeset by Tradespool Ltd, Frome, Somerset
Printed by St Edmundsbury Press Ltd, Bury St Edmunds, Suffolk

CONTENTS

INTRODUCTION

THE SCOPE OF THE COURSE

The course, in three parts, provides an introduction to the language, culture, and literature of the Romans. It covers the syllabus for the General Certificate of Secondary Education and the Scottish Certificate of Education (Standard Grade), except for the set books.

The first two parts take the form of a narrative which tells the story of the life of the poet Horace. Part I covers his boyhood in Venusia and his schooling in Rome; it ends with his departure from Rome to university in Athens. Part II covers his life in Athens as a student, his service in Brutus' army at Philippi, his return to Italy, his service as a clerk in the treasury, his friendship with Virgil and Maecenas, his life on the Sabine farm, and his relations with Augustus. It ends with the deaths of Maecenas and Horace.

Whereas Part I is largely fictional, Part II is based increasingly closely on historical sources and finds Horace near the centre of important events. In this part we try to build up a solid picture of the Augustan revolution.

Part III consists of extracts from authors of the late Republican and Augustan periods.

The sequence of grammar and syntax for Part I is set out on pages 7–8. The grammar comprises regular nouns and adjectives of all declensions, common pronouns, the active verb (all tenses of the indicative, imperatives, and present infinitive). The syntax comprises the simple sentence (statement, question and command) and temporal and causal clauses with the indicative. Part II, linguistically, covers most of the remaining basic grammar and in syntax most common constructions. Conditional clauses with the subjunctive, gerunds and gerundives, and relative with the subjunctive are left to Part III.

GCSE National Criteria (Classical Subjects) **2.1** states five aims for Linguistic Subjects:

1 to develop, at an appropriate level, a competence in the language studied;
2 to read, understand, appreciate and make a personal response to some of the literature in the original language;
3 to acquire some understanding of the civilization within which the literature studied was produced;
4 to encourage a sensitive and an analytical approach to language by seeing English in relation to a language of very different structure and by observing the influence of the ancient language on our own;
5 to encourage the ability to observe, abstract, and analyse information paying due regard to evidence and to develop a sympathetic awareness of others' motives and attitudes.

The aims which it will be hardest to achieve and which, we feel, justify the need for a new course are **2** 'to appreciate and make a personal response to some of the literature in the original language', and **5** 'to develop a sympathetic awareness of others' motives and attitudes'. These aims are unlikely to be achieved unless, from the very start, our pupils read Latin not as a linguistic jigsaw but as a vehicle for conveying meaning to which they habitually make a personal response. We have tried to construct a narrative which will evoke such a response, at first at a simple and banal level, but gradually becoming more sophisticated. As an understanding of Roman culture and history develops both through the Latin narrative and through the background sections which follow each chapter, and as Horace himself slowly emerges as a sympathetic character, pupils may find it possible to respond intelligently to the actual poems embedded in the later stages of the narrative.

These poems and extracts are introduced from Chapter 12 of Part II onwards; in testing they have proved unexpectedly successful. The reason for this seems to be that every poem occurs in a context understood by the pupils and arises naturally from the narrative, so that the poems appear to be the reaction of Horace to a known situation. And, in so far as Horace has emerged as a sympathetic character, pupils can and do form a personal response. Admittedly, the context foisted on the poems by the authors of the course will seldom bear scholarly scrutiny. The response made by our pupils will not be that made by an educated Roman or a contemporary scholar. But we have found that our pupils can cope with the questions we ask on the poems, which gradually become more demanding. This key topic will be dealt with more fully in the introduction to Part II.

The choice of Horace as our central character has certain other advantages. First, he is an exact contemporary of the younger Marcus Cicero, to whom, at the cost of some bare-faced historical fiction, we give a major role. This enables us to introduce his famous father, the one man about whom we know at first hand more even than we know about Horace, and we are able to use extracts from his letters in the narrative. Secondly, his friendship with Virgil enables us to prepare the way for the extracts of Virgil in Part III. By the time our pupils come to read extracts of Cicero, Caesar and Virgil, the authors will be old friends, or at least acquaintances, whose social and historical background is already partly known. The literature will not, we hope, seem remote and unreal but related to what they have already read and even relevant to their own experience. We believe that this treatment will make it possible to go some way towards fulfilling the aims laid down in the National Criteria, which appear truly formidable considering the age at which our pupils will be sitting the examination and the short time for which they will have been studying Latin.

LINGUISTIC PRINCIPLES

We have tried to combine what we consider the best features of both the modern and the traditional methods of teaching Latin.

From the modern method, exemplified by the Cambridge Latin Course, we accept firstly that the aim of any Latin course should be the acquisition of reading skill and that everything else, linguistically speaking, should be subordinate to this aim. Hence composition from English into Latin is used mainly as an adjunct, to practise grammatical forms and concepts; for this purpose we believe it has an important role in the early stages.

Secondly, we accept that the Latin language should be taught in a Roman context, so that understanding of the language and the culture proceed *pari passu*.

Thirdly, we accept that the acquisition of reading skill is partly an inductive process; that is to say, the student learns from experiencing the language as an instrument conveying meaning, not simply by analysis. As a broad principle we believe that the student should first read with understanding (and, if required, translate) and then study the grammar and syntax he has already met in context. We do not stick rigidly to this principle; if experience suggests that it is more helpful to do so, we explain grammar etc. before the narrative. But it remains true that in this course the first experience of new grammar and syntax (in the captions below the pictures which introduce each chapter in Parts I and II) always occurs before explanation.

Although we accept what we have called an 'inductive approach', we also believe firmly in the necessity of learning grammar and vocabulary thoroughly. We do not hold that 'immersion' in the language will enable students to form a 'personal' grammar, as we do in our native language. This may be possible in a modern language, given enough time, but it will not work in an ancient language in a strictly limited time. We tabulate grammar in the traditional form, though the order in which we introduce it is not traditional. For instance, the nominatives and accusatives of the first three declensions are introduced before any other cases are used (except for ablatives after prepositions, which are treated as lexical constants, requiring no explanation). This occasions no difficulty – accusative endings in the singular: **-am**, **-um**, **-em**; in the plural: **-ās**, **ōs**, **-ēs**. We can also introduce without comment accusatives of the fourth and fifth declensions, if they are necessary to the narrative.

Some teachers may be irritated, quite reasonably, by the introduction of some grammatical features which are not explained, e.g. in Chapter 1 **in** + ablative, not explained until Chapter 4. We must ask them to believe that this does not matter, as long as the Latin is understood (and can be translated) in context. By the end of Part I all the relevant grammar has been tabulated and learnt.

From the traditional method we accept the necessity and importance of teaching grammatical analysis. This is not an easy skill for the young to learn, but unless it is mastered in the early stages and becomes habitual and instinctive, trouble is stored up for later. The difficulty is increased by the fact that now formal grammar plays little or no part in the teaching of English. In the Pupils' Introduction we have tried to explain the names and functions of the basic parts of speech. We include exercises in analysis and suggest a notation for practising it.

PART I

THE LINGUISTIC CONTENT

Sequence of grammar and syntax

Chapter 1

Verbs: 3rd person singular, present indicative.
Nouns: 1st declension, nominative and accusative singular.
Subject, verb; subject, object, verb.

Chapter 2

Verbs: 3rd person singular, all conjugations.
Nominative and accusative singular, 2nd declension masculine.
Masculine and feminine gender.

Chapter 3

Verbs: **est** + complement.
Agreement of adjectives.

Chapter 4

Singular and plural.
Verbs: 3rd person plural, all conjugations; compound verbs.
Nouns: 1st and 2nd declensions, nominative and accusative plural.
Prepositions + accusative and ablative.

Chapter 5

Verbs: present tense, all conjugations, all persons.
Questions with **cūr?** and **quid?**
Adverbs.

Chapter 6

Verbs: imperatives singular and plural, all conjugations.
Nouns and adjectives: 3rd declension, nominative and accusative, singular and plural.

Chapter 7

Nouns: genitive case, 1st three declensions, singular and plural.

Chapter 8

Verbs: present infinitive, all conjugations.
Prolative infinitive; **nōlī**, **nōlīte**. **eō**.
The mixed conjugation.

Chapter 9

Subordinate clauses: **ubi**, **dum**, **quod**.
Verbs: **sum**, **adsum**, **possum**.

Chapter 10

Nouns: neuter nouns of the 2nd and 3rd declension.
Adjectives: **bonus**, **trīstis.**

Chapter 11

Nouns: dative case, first three declensions; indirect object.

Chapter 12

Verbs: compound verbs.
Nouns: revision of the first three declensions.
Pronouns: **ego**, **tū**, **ille**, **nōs**, **vōs**, **sē**.

Chapter 13

Verb: past tenses, imperfect and perfect.

Chapter 14

Numerals: one to ten.
Expressions of time and place.
Verbs: perfect and aorist meanings.

Chapter 15

Verbs: pluperfect tense.
Questions.

Chapter 16

Nouns: fourth declension.

Chapter 17

Verbs: the meaning of the imperfect tense; compound verbs.
Revision of adjectives.
Third declension adjectives.

Chapter 18

Nouns: uses of the ablative.
Verbs: dative verbs.
is and **ille**.

Chapter 19

Nouns: fifth declension.
Verbs: mixed conjugation.
Compounds of **dō**.
hic.

Chapter 20

Numerals: eleven to twenty.
Verbs: future and future perfect tenses.
sum and **possum** revised.

Verbs

The present tense only is used until Chapter 13, where the imperfect and the perfect (aorist) are introduced simultaneously. This makes for a certain amount of awkwardness in the narrative in places, but Latin authors are themselves so fond of the historic present that on the whole it works well enough. The purpose of this is to make students thoroughly familiar with the basic patterns of the Latin sentence while keeping the grammatical content as slim as possible. The future and future perfect (the rarest tenses in Latin) do not occur until the last chapter. On the other hand, imperatives (Chapter 6) and present infinitive (Chapter 8) are introduced early, since they are easily learnt and their introduction makes it possible to write a much more lively narrative. Since the passive voice is not learnt in Part I, neither the supine nor the past participle is given in principal parts; but a few past participles passive are introduced as ordinary adjectives (e.g. **parātus**, **territus**, **commōtus**), which will help when past participles are introduced in Part II.

Vocabulary

Each narrative is followed by a select vocabulary to be learnt; we have kept these lists as short as seemed possible in relation to our target. Words occurring in the vocabulary list of each chapter are not glossed in the narrative; other unknown words are glossed in the margin – a clumsy device but necessary for fluent reading. We do our best to reinforce common words by bringing them in repeatedly. When a word has occurred five times or more, we cease to gloss it, even if it has not occurred in the select vocabularies (such words can, of course, be looked up in the General Vocabulary, if necessary). We very occasionally leave an unknown word unglossed, if we consider its meaning quite clear from the context, e.g. **flamma** in a fire context. After considerable debate we decided not to gloss any of the pictures. We do not claim that the meaning of every word is clear from the pictures, and we draw attention in our chapter by chapter commentary to words which may cause trouble.

Exercises

Every chapter is followed by Latin sentences for translation, testing understanding of the grammar etc. which has just been learnt; sometimes we ask for a written analysis of these sentences, using a notation which we suggest. Most chapters include a few sentences for translation from English into Latin; these employ a limited vocabulary, since we consider that their sole purpose is to test understanding of and actively exercise the grammar and syntax just learnt. In many chapters there are completion exercises and at intervals exercises in word-building.

The last exercise of each chapter is perhaps the most important. This usually consists of two or three paragraphs of Latin developing or continuing the narrative. A translation of part of this is required; on the rest comprehension and grammar questions are asked. The comprehension questions are mostly on the sense and precise answers can be given. Pupils must understand the Latin as exactly as if they had been asked for a translation, but from Chapter 6 onwards there is usually at least one open-ended question, which either invites a personal response to the story or a judgement on the characters of those involved. These questions do not admit of precise answers and might be considered unsuitable for written exercises, but they play an important part in encouraging pupils to relate what they are reading to themselves and to begin to evaluate and to exercise elementary critical judgement. Attempts to answer such questions are highly relevant to the aims we discussed above and may gradually influence the way in which the main narrative is read. Finally, whereas the narratives are intended to be treated orally in class, the exercises are meant to be written.

PRONUNCIATION AND READING ALOUD

Throughout the course we consider it most important that Latin should be read aloud. Translation of the picture captions or the narrative should never be attempted until the teacher has read the Latin aloud to the class at least once, and pupils also should be given the chance of reading Latin aloud intelligently, as often as possible. A foreign language should be learnt through the ear as well as through the eye and this practice speeds up progress. It is so important that the Latin should be read correctly that we give an abbreviated note on the pronunciation of Latin here; we do not include this in the Pupils' Introduction, since it really must be taught orally.

Latin, unlike English, is spelt exactly as it is sounded and so correct pronunciation will make it easier both to recognize grammatical forms quickly and to spell accurately.

The Latin alphabet contains fifteen consonants (omitting the letters *x*, *y*, *z*, which are only found in words transliterated from Greek, and *k*, which occurs only in the word **Kalendae** and derivatives). It contains five vowels.

Consonants

Most of the consonants are pronounced, for practical purposes, as in English, except:

1. *i* consonantal, as in **iam**, is pronounced like English *y*.
2. *v* (= *u* consonantal), as in **vīdī**, is pronounced as English *w*.
3. *r* is rolled, as in Scottish, and is always sounded; e.g. **ars** – both **r** and **s** are sounded.
4. *c* is always hard as in *cat*.
5. *s* is always pronounced as in *sit*, never as in *rose*.
6. *g* is always hard as in *God*, except when followed by *n*; *gn* is pronounced *ngn* as in *hangnail*.
7. *h* is always sounded as in English *hope*.
8. *q* is never found except in the combination *qu*, sounded as in English.

Vowels

The five vowels each have a short and a long version:

1. *a* short, as in English *cup* (NB not as in *cap*).
2. *a* long, as in English *father*.
3. *e* short, as in English *pet*.
4. *e* long, as in French *gai*.
5. *i* short, as in English *dip*.
6. *i* long, as in English *deep*.
7. *o* short, as in English *pot*.
8. *o* long, as in French *beau*.
9. *u* short, as in English *put*.
10. *u* long, as in English *fool*.

Diphthongs

1. *ae* as in English *high*.
2. *au* as in English *how*.
3. *ei* as in English *day*.
4. *eu* e-u (occurs rarely, not properly a diphthong).
5. *oe* as in English *boy*.
6. *ui* u-i.

Throughout the book, naturally long vowels are marked with a macron, except those which occur in an initial capital letter, e.g. the first *i* in **Italia** is long but is not marked with a macron; apart from this all vowels not so marked are short.

THE NARRATIVE

Although the narrative centres round the life of Horace, Part I contains a digression of five chapters in which the schoolmaster tells the stories of part of the *Iliad* and the first part of the *Aeneid*. We make no apology for this; the stories are basic to all Latin and, indeed, to all European literature, and it seems to us that any course intending to provide an introduction to the classics must include them. The stories are inevitably told in a jejune and skeletal form, since at this stage the linguistic resources are minimal. Teachers may like to add flesh to these bare bones by reading their class translations of some of the originals.

We know very little about Horace's early life. He says a good deal about his father (*Satires* 1,6) but he never mentions his mother and we do not know whether he had any brothers or sisters. We have invented a sister for him, Horatia, and throughout Part I we have resorted to fiction. Our running commentary will draw attention to the historical basis of these fictions, as they occur. In a few places we consciously depart from historical fact; for example, we make Quintus meet the younger Marcus Cicero at Orbilius's school. In fact, Cicero educated his son at home and it is unlikely that he ever went to this school.

THE BACKGROUND MATERIAL

The background material which is placed at the end of every chapter is intended bit by bit to build up a rounded (but incomplete) picture of Rome of the first century BC. We hope that it will meet the GCSE aims **3** ('to acquire some understanding of the civilization within which the literature studied was produced') and **5** ('to encourage the ability to observe, abstract, and analyse information paying due regard to evidence and to develop a sympathetic awareness of others' motives and attitudes').

We have followed each passage with a question which we hope will stimulate further thought about the topics raised, especially in the matter of how the civilization of Horace's Rome relates to the modern world. (Here differences are as important as similarities.) Our questions can usually be answered on the basis of the background passages, the illustrations, the Latin story and, of course, the pupils' own experience. The level of sophistication demanded by these questions varies, and you may think it best to omit some if they are too naïve or too difficult for your pupils.

Where a topic appears to have struck a particular chord, further reading should be encouraged. Some books recommended for following up topics raised are:

J. P. V. D. Balsdon: *Life and Leisure in Ancient Rome*, Bodley Head.
John Boardman, Jasper Griffin, Oswyn Murray (edd.): *The Oxford History of the Classical World*, Oxford.
Jérôme Carcopino: *Daily Life in Ancient Rome*, Penguin.
O. A. W. Dilke: *The Ancient Romans, How they lived and worked*, David and Charles.
U. E. Paoli: *Rome, Its People, Life and Customs*, Longman.
David Taylor: *Cicero and Rome*, Macmillan.
Moses Hadas: *Imperial Rome*, Time-Life Books.
G. I. F. Tingay and J. Badcock: *These Were the Romans*, Hulton Educational.

The narrative starts in the country, or rather in a remote country town, which enables us to begin with very simple accounts of ordinary Roman life. As we move towards a more historical context, we have had to be highly selective in what we include.

The background material can either be read out loud in the classroom by the pupils and/or the teacher or set for homework. Discussion of the issues raised should be encouraged. You might try the occasional debate.

SOME SUGGESTIONS ON METHOD

It is intended that the narratives should be treated orally and taken fast. The appropriate speed of reading (and translating) the Latin can only be gauged by the teacher in relation to the ability of his class, but the stories are often long and, if they are taken too slowly, boredom will result. Moreover, we wish to inculcate the habit of fluent reading and this will grow, if the class is stretched to the limit of its ability.

We have already stressed the importance of reading Latin aloud. We are prepared to assert dogmatically that pupils should never attempt to translate either the captions below the pictures or the stories themselves until the Latin has been read aloud by the teacher at least once, and that the minimum reading unit should be the complete sentence, up to the nearest full stop, e.g. the opening sentence of Chapter 1 reads: **Scintilla in culīnā labōrat**; **cēnam parat**. We have kept the paragraphs short so that, as soon as some fluency has been established, the paragraph may become the usual unit of meaning. In early stages, and later in difficult passages, it may be necessary to read the Latin aloud more than once.

Understanding may be tested either by asking pupils for a straight translation, or by asking questions in English about what is happening (comprehension questions). There is, however, no magic in reading Latin aloud and at times there will undoubtedly be failures of understanding, which can be cleared up by asking for an analysis of the difficult passage.

It may be remarked here that understanding a foreign language and translating it are different skills. When we say that the overriding aim of the course is to teach our pupils to read Latin fluently and intelligently, we mean ideally that we want them to understand Latin as they read it without translating. Translation is the traditional and the most precise method of testing understanding. Teachers will certainly use if for much of the time, but, especially as our stories are rather long, it is highly desirable to vary the reading lesson by sometimes asking comprehension questions instead of demanding a full translation. The ultimate test of understanding the sense of a passage is for the teacher to read it aloud in Latin while the pupils do not look at the text and then test their understanding by comprehension questions or by asking them to retell the story in their own words in English. As an occasional variant this exercise is much enjoyed.

Translation is a complex process, especially from an ancient language such as Latin, in which structure and idiom are so different from our native tongue. The acceptable translation must not only be accurate but must express the meaning in convincing English; even at a very early stage pupils will be choosing, unconsciously, between better and worse versions; for instance, take the sentence 'When the dog barks, the hare is afraid (**timet**)'; 'the hare fears' would be unacceptable. If from the start our pupils are trying to elicit meaning from intelligible units (whole sentences or paragraphs), they are more likely to be successful in their English versions.

We advocate speed and fluency in reading and realize that such a method might be thought to result in hazy and inaccurate understanding. We do not think that this is so but, if it were, the exercises, which require close examination of every word and strict analysis, would act as a corrective.

COMMENTARIES ON EACH CHAPTER

⋆ *denotes illustration.*

⋆ Cover: this still life with eggs and thrushes from Pompeii was painted between about 55 and 79 AD. It is in the National Museum at Naples.

⋆ Title page: this silver statuette of a shepherd carrying a sheep in a bag probably dates from the first century AD. It is in the British Museum.

Chapter 1

This chapter places Horace in a geographical and family context. We know that his full name was Quintus Horatius Flaccus and that he was born in Venusia on 8 December 65 BC. The facts are given in the *Vīta Horātiī*, derived from Suetonius's *Dē Virīs Illūstribus*, which begins: 'Quīntus Horātius Flaccus Venusīnus . . .' and which gives his date of birth in the last paragraph.

Venusia, modern Venosa, was one of the more important towns in South-East Italy, lying on the Via Appia, which led from Rome to Brundisium. A colony of Roman citizens was settled there at the end of the Third Samnite War (291 BC) and Sulla made another settlement of veterans not long before Horace's birth. It was a large town by Roman standards, but no doubt a high proportion of its population were farmers, like Horace's father, who is described by Horace in *Satires* 1, 6 as 'macrō pauper agellō' (a poor man with a meagre farm).

We know nothing about his mother, whom he does not mention, nor about any other members of his family. We have invented a sister, Horatia. His father, for whom Horace felt great admiration and affection, was, as we explain in the background section, a freedman (**lībertus**), i.e. he had been born a slave and would probably have bought his freedom from his master. A **lībertus** had full citizen rights except that he could not stand for political office; his sons were full citizens in all respects. A freedman usually took his master's names, and so Flaccus's master was presumably called Horatius Flaccus.

Treatment of the pictures and the captions

These are intended to be self-explanatory and various elements are not explained: **est**; the demonstrative **hic**, **haec**; the meaning of these should be made clear by gesture; so also **ecce!** and the whole phrase **in hāc pictūrā**. Successful treatment depends upon intelligent reading aloud (in Latin) with, if necessary, exaggeration of tone and gesture. If pupils are stuck, ask questions which elicit the meaning: 'Where does Quintus live?' 'Where is Apulia?' 'Where is Scintilla? What is she doing?', etc. After this an oral translation may be demanded;

occasionally it may be necessary to give a word, e.g. **iuvat** under the last picture.

Quintus is hungry

The narratives, as always, should be treated orally and translation should not be attempted until the Latin has been read aloud as often as necessary. After the first halting translation or question and answer session, the Latin should be read aloud again in whole paragraphs and an oral translation of each complete paragraph should be asked for; this will help to instil the habit of fluent reading. The vocabulary should be learnt as soon as the story has been dealt with.

Grammar: subject, object, verb

Unfortunately, more grammatical explanation is necessary in the first chapter than almost anywhere else in the book. This is because we cannot discuss grammar without being able to use grammatical terms. The terms used here are: sentence, verb, noun, subject, object, case. These have all been explained in the Pupils' Introduction, but it is unlikely that the explanations will have been fully understood. Teachers will no doubt have their own methods of making this stiff dose of grammar palatable; time spent on some analysis of simple English sentences would not be wasted.

Teachers should not despair if pupils are still hazy on analysis and grammatical terminology at the end of this chapter. Subsequent chapters and practice will gradually clear up difficulties.

Two other points are liable to give difficulty and will require more explanation than we have given:

1 suppression of pronoun subjects, e.g. **labōrat** *he* works;
2 **labōrat** (a) he works (b) he is working.

If pupils look for a one for one correspondence between words in English and Latin, they may be floored; but if from the start they are trying to grasp the meaning of whole sentences in a context, they are likely to come up with the right answer without reflection.

Background

It might be a good idea to discuss the concept of slavery before the pupils write the life of a slave or freedman. Slavery was a fundamental fact in the ancient world and it is worth remarking that in England it was not finally abolished until William Wilberforce's Act of 1807.

For the good treatment of slaves, see Seneca, *Letter* 47, and Pliny, *Letters* 8, 16. For the results of bad treatment see Pliny, *Letters* 3, 14.

★ pp. 11 and 12: we show a nobleman and two freedmen. The statue of the nobleman, dressed in a fine toga, is from the late first century BC. The two freedmen were formerly slaves of P. Licinius. At the top (in the pediment) are blacksmiths' tools: hammer, anvil and tongs. To the right are carpenters' tools: a drill with the bow used to rotate it, a marking knife, an adze and a short-bladed chisel. To the left are the **fascēs**, the axe and the rod used in the ceremony of freeing the slave.

★ p. 12: inscription (a) is from near Philippi (Barrow 160); (b) is from Ravenna (Dessau 1980).

Chapter 2

The pictures and captions should give little trouble, except perhaps for **arat** (since ploughing with oxen is so different from modern methods) and **in agrō** (since it might not be clear from the picture that this is a field).

Grammar: gender and verbs

1 The concept of gender should be readily understood by those who have done some French (neuters are postponed until Chapter 10).
2 Verb endings: **-t** is the ending for the third person singular for all active tenses of all verbs. Verbs of all conjugations are now used freely. It is better to gloss over the differences in stem vowels at present, if possible. The point is cleared up in Chapter 5.

Background

You may have to explain what a small-holder is (second paragraph). The words 'dictator' and 'dictatorship' in the fourth paragraph may also need explanation.

The 'vast country estates' referred in the fifth paragraph probably consisted of a number of small estates, individually not significantly larger than the farms worked by small-holders like Horace's father. The **colōnus** system had proved its worth. See background material for

Chapter 4.
p.17: the quotation in the final paragraph about three types of farm equipment is from Varro, *Dē Rē Rūst.* 1, 17, 1, 5–7.

★ p.17: in this section of the Travaux Champêtre mosaic from Cherchelle (Algeria), the man to the left is ploughing while the man to the right is sowing.

Chapter 3

The captions contain no vocabulary problems.

Grammar

The verb 'to be' is at present used only as a copula. Its other use (= exists) is explained in Chapter 11. The agreement in case and gender of subject and complement may need stressing before Exercise 3.3 is done.

Quintus is lost

The story is based on *Odes* 3, 4, 9–20, where Horace says that as a child he wandered from the home of his nurse Pullia (if this is the correct reading; we have not introduced her, since her very existence is suspect) on Mount Vultur, about five miles from Venusia, and worn out by play he slept in a wood. The doves covered him with leaves, like the Babes in the Wood. He attributes this and his miraculous escape from the attention of bears and vipers to the protection of the gods. Unfortunately, these graphic, if fabulous, details proved to be beyond the linguistic scope of this chapter.
p.18 **nōn redit malus canis**: subject following verb, introduced for the first time, may cause trouble, especially as the third declension has not yet been learnt.

Background

★ p.21: in this relief from the first century BC, the wife sits in a chair with her baby on her knee while her husband reclines. The older children stand on either side. (Museum Calvet, Avignon.)

We have made use of the idealized picture of the chaste Roman wife in Virgil, *Aeneid* 8, 408–13. There is, of course, particularly rich ground for discussion here. How different is the role played by women in the twentieth century? A useful book here is L. P. Wilkinson's *Classical Attitudes to Modern Issues* (London: Kimber, 1978).

The inscription is *CIL* 1, 1007. Compare this epitaph from the second century BC:

> Stranger, what I say is not long. Stand and read it. This is the unlovely tomb of a lovely woman. Her parents called her Claudia by name. She loved her husband with all her heart. She had two sons: one of them she leaves on earth; the other she has placed beneath the earth. Her conversation was delightful, her deportment graceful. She looked after the home. She spun wool. I have spoken. Go on your way.

Chapter 4

The captions contain no vocabulary problems.

Grammar: singular and plural

This concept presents no difficulty, since changes to noun endings are made in English and French to indicate number.

Argus saves Horatia

This chapter centres around the river Aufidus, modern Ofanto, one of the largest rivers in Southern Italy, about ten miles north of Venusia (Argus led them quite a dance). Horace mentions it several times; he describes himself as born near the far-sounding Aufidus ('longē sonantem nātus ad Aufidum', *Odes* 4, 9, 2) and says that he will be remembered 'where the raging Aufidus roars' ('quā violēns obstrepit Aufidus', *Odes* 3, 30, 10). Clearly the river was not, and is not, safe for bathing; Quintus and Horatia were very rash and Argus did a good job.

Grammar: prepositions + ablative

The ablative case is used only after prepositions until Chapter 17 where other uses are explained.

Exercise 4.6

This type of exercise is more instructive than the traditional question. 'Give one Latin word derived from . . .'. You may like to make up similar exercises in subsequent chapters, but remember that the past participle of verbs has not been introduced.

Grammar: compound verbs

This useful information is placed early in the course. Pupils should be expected to recognize that e.g. if **mittō** means 'I send', **ēmittō** means 'I send out'. We have not thought it necessary to include all compound verbs in vocabularies or to gloss them.

★ p.28: this shows women talking at a fountain house. Two of them are returning from the fountain with their full jars upright on their heads; two more are approaching with their jars carried sideways. The fifth is filling her jar from the lion-headed spout. The scene is from a water jar made *c*.520–500 BC in Athens; hence the Greek lettering.

Background

★ p.29: this shows the colonnade round the Forum at Pompeii. The view is north towards Vesuvius.

We hope that the meaning of the word 'colony' in a Roman context has been made clear in the text. In Britain there were colonies at Colchester, Lincoln, Gloucester, and York.

It may be necessary to say a bit more than we have about the word 'consul'. But avoid going into too much detail at this stage.

★ p.30: this hill town is Barrea L'Aquila in the Abruzzi.

p.30: the election poster: Dessau 6438d. Among the thousands of notices painted on the walls of Pompeii are many election posters, some painted by party factions, e.g. 'Innkeepers, vote for Sallustius Capito!'

Chapter 5

Captions: **cūr**? and **quod** may need explanation.

Grammar: verbs

The complete present tenses of all four conjugations are introduced simultaneously. This presents no difficulty if the person endings are learnt first. We have set out four verbs with the vowel included in the stem, which is more correct than the traditional format and quicker to learn. But

1 **amō**: the **a** disappears through contraction from **amaō**;

2 consonant stems (and **-u** stems) insert vowels before the person endings (for euphony); strictly speaking these are not the same as in the fourth conjugation; **i** is short in the third and long in the fourth conjugation. The third person plural of these conjugations (**-unt**) always gives trouble when pupils write Latin.

The school of Flavius

Horace (*Satires* 1, 6, 72–6) says that his father 'refused to send me to the school of Flavius, where the big boys, sons of big centurions, went with their satchels and tablets hung over their left shoulders, taking their eight asses on the Ides of each month, but he dared to take me to Rome...'.

We take the liberty of making Flaccus send Quintus to Flavius's school but he removes him after a riot (Chapter 13). The local school was cheap; the boys took the fee to the schoolmaster themselves and they had no tutor to carry their satchels and tablets. The big boys would be the sons of Sulla's veterans, centurions, who were now the big men of Venusia. Horace in later life can look back on the school with characteristic irony.

p.33 **exspectā mē**: imperatives creep into this chapter. The first is glossed. In the other examples sense and punctuation should help pupils over the linguistic hurdle (**manēte! venīte!** etc.). Imperatives are explained fully in the next chapter.

★ p.34: in this marble relief from Neumagen (in Gaul), of about 200 AD, two boys sit on **sellae** with papyrus **volūmina** unrolled; a third arrives late, carrying his **capsula**.

Background

para. 1: the difference between Italian dialects: in one of them (Oscan) the word for five (**quīnque** in Latin) was **pompe**.

para. 5: the Roman poet who complained at being woken up by the noise of the schoolmaster was Martial (*Epigrams* 9, 68): 'What have you to do with us, you cursed schoolmaster, a man hateful to boys and girls alike? The crested cocks have not yet broken the silence of night: already you are making a noise with your cruel voice and your thwacks.'

★ p.37: this is a marble relief of the second century AD in the Louvre Museum. From left to right: the father holds his child, accepting him into the family; the boy at play rides a toy chariot drawn by a donkey; he receives instruction.

Chapter 6

Captions: **stant, prope, tacēte** are new words; the pictures may make their meaning clear.

Grammar: imperatives

With the use of imperatives, vocatives appear. As this case is the same as the nominative for all nouns except for the singular of 2nd declension nouns in **-us**, we never list vocatives in any summary of grammar. The second caption gives our first example of vocative in **-e, Quīnte**. We suggest a passing comment, but it gives no difficulty except where it occurs in translating from English into Latin. **fīlī** and **mī fīlī** etc. are glossed when they first occur.

Flavius's story

This chapter contains the longest narrative so far. It falls into two parts, the introduction (**fābula**), exercising imperatives, and the Death of Hector. The introduction should go very fast since the material is mostly a rehash of the previous story.

Grammar: the third declension

The accusative singular gives no trouble, since it only represents a change of vowel: 1st declension **-am**, 2nd declension **-um**, 3rd declension **-em**; but the plural, where both nominative and accusative are **-ēs**, presents genuine ambiguity and confusion. This can only be solved by clinging to the meaning in context.

If pupils ask why there should be so many declensions, the answer is that each declension has a different stem ending; 1st **-a**, 2nd **-o** (though this is obscured in classical Latin), 3rd consonant and **-i** stems (**nāvis** originally declined in the singular **nāvis, nāvim, nāvis, nāvī, nāvī**); 4th **-u**, 5th **-e**.

The Death of Hector

It has been impossible to keep genitives out of this chapter; they are glossed as they occur. The genitive is learnt in the next chapter.

★ p.43: this is from a vase of about 490 BC. Achilles and Hector carry the arms of Greek hoplites of the fifth century BC, not of Homeric warriors.

Background

You may think that the Homeric question is too much to take on board at this stage. If so, omit paragraph 1. The most amusing account of the Judgement of Paris – to which our own version owes something – is that of Lucian (*Deōrum Dialogī*, 20). This has been entertainingly translated by Paul Turner: *Lucian: Satirical Sketches* (Penguin Classics, 1961), pp.55–64.

The reference to Helen's face which 'launched a thousand ships' is to the scene in Christopher Marlowe's *Doctor Faustus* where Faustus addresses the phantom Helen of Troy:

> Was this the face that launched a thousand ships,
> And burnt the topless towers of Ilium?
> Sweet Helen, make me immortal with a kiss . . .
> I will be Paris, and for love of thee
> Instead of Troy shall Wittenberg be sacked . . .

★ p.48: the votive relief of Minerva was carved at Athens between 470 and 450 BC. The relief of Neptune, Apollo and Diana is from the frieze of the Parthenon, built 447–38 BC.

Chapter 7

Captions: **rapit, tabulās, pūpās** all are new, but the pictures should make their meaning clear.

Grammar: genitive case

The concept of the genitive gives no great trouble. It is worth pointing out that the genitive can either precede or follow the noun it depends on, after you have read and translated the captions.

The difficulty, which is bound to cause confusion sooner or later, is that **puellae** and **puerī** can be either nominative plural or genitive singular. There is no way of escaping this amphisbema. When in doubt, the only safe guide is the sense demanded by the context.

The basic use of the genitive here introduced is the possessive, but we have also allowed ourselves some partitive genitives, e.g. **multī puerōrum**. As this usage is similar to English 'many of the boys', it is better not to discuss the difference between these two uses.

The Fall of Troy

The story of the Fall of Troy is taken from *Aeneid* 2 (not from the *Iliad*, which ends with the burial of Hector).

★ p.51: part of a relief on a funerary urn of the late seventh century BC. The horse is mounted on wheels. The Greek warriors are clearly visible, some of them even holding their armour out of the apertures!

★ p.52: this Athenian black figure vase of about 550 BC shows Hecuba, the wife of Priam, stretching out her arm as she tries to stop Neoptolemus killing her husband. (British Museum)

Background

The story is taken from *Iliad* 24.

★ p.55 the ransom of Hector: this Athenian cup of about 490–80 BC gives a rather different view of the story from that of Homer. Priam (left) approaches an arrogant Achilles who reclines on a couch above the body of Hector. In Homer's version, Achilles takes great care that Priam and he should not be together with Hector's corpse. He feels this would arouse emotions dangerously. (Kunsthistorisches Museum, Vienna)

p.55 Achilles' heel, last para.: Achilles' mother dipped him as a child in the waters of the Styx to make him invulnerable. His heel, which she was holding him by, escaped the treatment.

Chapter 8

Captions: **iubet, cupiunt, nōlīte, dēbētis, cōnstituit** are all unknown and help will be required.

Grammar: the infinitive

The prolative infinitive is similar to English usage and gives no trouble.

Pupils may be surprised by **nōlī, nōlīte** + infinitive = don't. Teachers may wish to explain that these forms are imperatives and mean literally 'be unwilling to'.

Grammar: the mixed conjugation

Now that the infinitive has been introduced, it is no longer possible to disguise the identity of verbs of the mixed conjugation (so far treated as if they were 4th conjugation verbs). In translation and comprehension, these verbs cause no difficulty but in writing Latin they probably will.

The Wanderings of Aeneas

This chapter is drawn from *Aeneid* 3. In stories taken from the *Aeneid* we often have reminiscences of Virgil's actual words, which give the language a rather poetical colouring.
p.58 **īnsula Dēlos**: the teacher may need to explain that Delos was Apollo's birthplace and one of his main cult centres. Delphi was more important in classical times; Quintus visits Delphi and Delos in Part II. The whole conception of prophecy and consulting oracles is so unfamiliar to us today that some explanation is probably here called for. Pilgrims, both individuals and state embassies, came from all over the Graeco-Roman world and beyond to the cult centres of Apollo, the god of prophecy, especially to Delphi, throughout ancient times until the oracle was plundered and destroyed by the Emperor Arcadius in 398 AD. They came to ask not so much 'What will happen?' as 'What should I do?' Apollo gave his responses through a priestess who was possessed by the god. The oracles were often hard to interpret or ambiguous, like the one given on this occasion to Aeneas, which his wise old father, Anchises, has to interpret.
p.59 **festīnant . . . ascendere**: they hasten to board (good classical Latin).

Vocabulary

Vocabulary lists are now given with verbs in the infinitive and no conjugation stated. Stress that 2nd conjugation verbs have long **ē** in the infinitive, 3rd short **e**, and that **cupiō** and **accipiō** are mixed conjugation. Vocabulary lists should always be read aloud by the class as well as by the teachers.

Background

This should probably be read before the Latin story in this chapter. Homer's telling of the stories of Scylla and Charybdis and the Cyclops is, of course, inimitable. If you have a translation of the *Odyssey* readily available (especially E. V. Rieu's for Penguin), you might read from Book 9 or 12. We have given only enough information about Scylla and Charybdis to make our Latin story comprehensible.

★ p.63: this lively picture is from a Greek vase in the British Museum.

★ p.64: from an Athenian vase of about 480 BC. The Cyclops, holding back the great rock at the entrance to his cave, lets out his flock. Odysseus and one of his companions cling on beneath their bellies. (Hunt Collection, Fort Worth)

Chapter 9

Captions: **dum** will need explanation. **ventī**, **tempestās**, **undīs**: meanings should be clear from the pictures.

No new grammatical forms are introduced in this chapter.

Wrecked in Libya

p.66 **trīstis**: the adjective is used adverbially; this is very common in Latin with certain adjectives. The translation 'sadly' should be accepted; compare **vōs laeta accipiō**, p.67.

★ p.66: this is not from the ancient world. It appears in an Italian manuscript of the *Aeneid*, written and illuminated in the late fifteenth century. We see (from left to right) the Trojans landing in Libya, Venus showing Aeneas and his friend Achates a good omen of flying swans, and Dido welcoming Aeneas. (British Museum)

Grammar: subordinate clauses

The ability to recognize subordinate clauses and bracket them off is one key to success in reading real Latin, especially Latin prose. It is very important that this skill should be acquired at an early stage, when the Latin is still very simple. The terminology is tiresome but the technique is not difficult to learn.

Exercise 9.6

You should tell your pupils that in this passage Aeneas is speaking. Rather a long story for written translation but the first part has already been told and so should go quickly. The first paragraph might be done orally. The story is taken from *Aeneid* 2 and is told more fully in the background section.

★ p.70: in this carved gemstone, now in the British Museum, Aeneas carries on his shoulder his father – who holds a lustral bowl – and leads his son by the hand as he escapes from Troy. He provides an archetype of **pietās**, fulfilling his duty to his religion, his family and his country (by ensuring the survival of the next generation and a new ruler). The motif became a significant feature of Augustan propaganda. There appears to have been a famous statue-group representing the subject in Augustus's Forum.

★ p.70: a Roman mosaic of the third century AD, now in Tunis.

Chapter 10

Captions: **nūntium**, **lītus**, **imperia**, **nūntiat**: all of these may perhaps be 'guessed'.

Grammar: neuter gender

The concept of neuter gender may cause difficulty; it would not do so, if all things were neuter, but why should **hasta** be feminine, **gladius** masculine, and **tēlum** neuter? Pupils must accept this as a peculiar quirk of the Latin language.

Infēlīx Dīdō

This narrative is taken from *Aeneid* 4.

★ p.75: this picture is one scene from a mosaic telling the story of the love of Dido and Aeneas, from a Roman villa at Low Ham in Somerset (fourth century AD).

Exercise 10.5

The practice of holding funeral feasts and funeral games (*Aeneid* 5) in honour of the dead was common in ancient times but will seem strange to our pupils; perhaps they should be warned of what is going to happen in this passage.

Background

A summary of *Aeneid* 6.

We preserve Virgil's formulaic repetition which links Aeneas's meetings with the ghosts of his wife and Anchises (see p.71). Aeneas and the Sibyl leave the Underworld by the Gate of False Dreams. Virgil may be raising an element of doubt about the destiny of Aeneas, which has already destroyed Dido. Is Aeneas in the right?

★ p.79: our illustration shows Cerberus after Hercules had kidnapped him from the Underworld as one of the labours set him by King Eurystheus. The terrified Eurystheus is seeking refuge in an urn. The vase was made in Etruria in the late sixth century BC.

Chapter 11

Captions: **flōrēs**, **dat**, **reddit**, **ostendit**: all may, perhaps, be 'guessed'.

Grammar: the dative case

The indirect object is not a difficult concept, but English confuses the issue by saying, 'I give you flowers', 'I tell him this', where 'you' and 'him' look superficially like direct objects. Analysis gives the correct answer; pupils should be warned and practised. In Exercise 11.1 you should ask for two translations of each sentence, e.g. The daughter gives flowers to her mother = the daughter gives her mother flowers.

The only use of the dative which is explained in the pupils' text is the indirect object but we allow some examples of the dative of interest to appear. The first is in the second paragraph of the story, when Scintilla says: **ego cibum parō tibi patrīque. tibi** is glossed but some passing explanation may be required, e.g. 'the dative sometimes means "for" rather than "to"'. The dative of interest will be explained in Part II and is glossed wherever it occurs in Part I.

Horātia Argum servat

We return at last to the main narrative and give Horatia a crack of the whip. The story is rather long but should give little trouble.

Vocabulary

Stress **dō, dăre** – the only first conjugation verb to have a short **a** in the infinitive. **ferō**: no infinitive given because it is irregular (we gloss it where it occurs).

Exercise 11.6

A variation on the usual translation exercise; the sentences form a continuous silly story.

Background

We summarize the last six books of the *Aeneid* in half a paragraph. You may wish to expand on this!

'Romulus and Remus' is the first of several stories from Livy, Book I. We aim to give an impression of Roman traditions and values, as well as a skeletal outline of Roman history up to the time of Horace.

Romulus and Remus were the children of Mars. It is worth asking what this tells us about the basic nature of the Romans. This will help with the questions at the end of Chapter 13.

Six Vestal Virgins were chosen by the Pontifex Maximus to serve the goddess Vesta and tend the sacred fire. They were selected from girls aged between six and ten and served for thirty years, after which they might marry. If they were proved unchaste during their period of service, they were entombed alive.

★ p.87: this statue group, with its personification of the protective River Tiber, is from Hadrian's villa at Tivoli (second century AD).

★ p.87: the 'Capitoline Wolf' is an Etruscan bronze from the late sixth or early fifth century BC. It used to stand on the Capitol. The twins were added in the fifteenth century AD. (Capitoline Museum, Rome)

Chapter 12

Captions: **sē lavat, exercet** may be 'guessed'.

Quīntus mīlitēs spectat

★ p.89: in this **larārium** from the house of the Vettii at Pompeii, the **paterfamilias** is in the centre, making an offering to the **Larēs** who stand on either side of him holding drinking horns. (Alternatively, the man in the toga may represent the *genius* of the family.) Beneath, the sacred serpent is about to consume an offering laid out for it. The spirits of the dead were thought to make their appearance in the form of serpents. (Since serpents live in holes in the ground, they have since very early times been thought of as vehicles for the spirits of the dead.)

The pupils should read the background section on Roman religion before the Latin story is read. Family worship of the **Larēs**, **Penātēs**, and **Vesta** was a regular feature of the old Roman religion. The **Larēs** were the deified spirits of dead ancestors; the family used to pray to them every morning and offer gifts, such as incense, wine or flowers (cf. Plautus, *Aulāria*, Prologue). The **Penātēs** were the spirits that watched over the larder. Prayers were made to Vesta, goddess of the hearth, before the main meal every day. So the religion of the home and family played an important part in the life of the ordinary Roman, especially in the country, where the old traditions were kept up. In fact, these cults and the ancient agricultural cults probably meant more to the Roman countryman than the worship of the Olympian gods, who were either imported from Greece or so assimilated to their Greek counterparts that they ceased to have much resemblance to the native Roman spirits with which they were equated.

The festival of the Parīlia was celebrated on 21 April in honour of Pales, protector of the flocks. By Horace's time the celebrants could not say whether Pales was one or two spirits nor whether he/she/they was male or female. No doubt the cult went back to a time when Roman religion was not anthropomorphic. It is described in Tibullus 2, 5, 78ff. and Ovid, *Fasti* 4, 721–862.

The section on Crassus might be read halfway through the story, when Quintus and Gaius slip away to watch the soldiers. There is more about Crassus and the first Triumvirate in the background sections to Chapters 17 and 19.

Grammar: pronouns

1 Subject pronouns used for emphasis. English gets the same effect in speech by stressing the pronoun, and in writing underlining is sometimes used. Latin regularly uses **ille** to point a change of subject.
2 Reflexive pronouns. We have used the French reflexive verb as a model and have not thought it necessary to point out that in both French and Latin **mē**, **tē**, **nōs** (nous), **vōs** (vous) are used as both reflexive and non-reflexive pronouns, whereas **sē** is used only reflexively and has no nominative. Teachers may need to clarify this point, if questions are asked.

Latin often uses transitive verbs with the reflexive pronoun, e.g. **mē vertō** = I turn (myself), whereas English uses the same verb both transitively and intransitively. Our examples make this clear but not explicit. Difficulty may arise in translating from English into Latin, e.g. Exercise 12.5, 'The old man turns . . .'. **vertit** will not do; it must be **sē vertit**.

Exercise 12.6

exercentēs is not glossed (present participles are not taught until Part II but have been used so frequently that in context pupils will probably give the correct answer; if questions are asked, they should be treated as adjectives).

★ pp.96–7: this frieze, from a monument of the early first century BC, shows a purification being performed with the sacrifice of a bull, a sheep, and a pig. The monument is known as the altar of Domitius Ahenobarbus and it is in the Louvre Museum.

Chapter 13

Captions: **ōlim** and **subitō** will require explanation.

Grammar: past tenses of the verb

Imperfect and perfect (aorist) tenses are introduced simultaneously. The grammar as set out looks formidable. It is worth stressing that there is only one set of endings for the imperfect tenses of all verbs (except **eram**) and only one set for the perfect of all verbs.

We have called the undifferentiated past tense 'perfect', as Kennedy does in his grammar, but we have only used the aoristic meaning in this chapter, which is, of course, far commoner than the true perfect. The true perfect with 'have' is introduced in Chapter 14. The meanings of the imperfect tense are further explained and exercised in Chapter 17. Most of the imperfects in this chapter may be translated 'were coming', etc. But several will be correctly translated by English simple past, e.g. **per tōtam hiemem Quīntus domī manēbat** Quintus stayed at home the whole winter. The difference between English and Latin usage cannot be glossed over and should be faced as soon as it occurs; Latin uses the imperfect tense for any action which is continuous, repeated or incomplete.

Tumultus

The narrative begins with a rehash of Chapter 5 but now the past tenses are used, so things are liable to start quite smoothly. Third persons, singular and plural, are used until half way through, when second persons begin to come in. Although the grammatical leap of two new tenses appears formidable, it will be found to work quite well.

Towards the end of the story Flaccus decides to take a job as an auctioneer's agent – his main task would have been to collect payments due – to swell his income. In a moving tribute to his father Horace tells us that Flaccus did this work; the passage is worth quoting in full (*Satires* 1, 6, 75ff.):

> My father dared to take me to Rome as a boy, to be taught the arts which any senator or knight would have his sons taught . . . nor was he afraid that someone might hold it against him if I followed a modest profession as an auctioneer or auctioneer's agent, as he was himself; nor would I have complained. For this I owe him all the greater praise and thanks. I would never be ashamed of such a father while I'm in my right mind.

Background

p.104, para. 1: the Etruscans had dominated the centre of Italy before the Romans took over. Their civilization, which flourished between the tenth and third centuries BC, was a remarkable one. They were both creative and practical and had a great influence on the early Romans, as well as supplying some of their kings.

★ p.105: this bust, popularly identified with the Brutus who drove out the tyrants, is from the late fourth century BC.

Chapter 14

Captions: **iter** must be given (with its gender); **contendunt** is new; **possumus, satis** and **prīmā lūce** must be given.

Quīntus Rōmam advenit

We make Flaccus and Quintus move at a fair pace, covering about thirty miles a day on their journey of 250 miles. Travel in Italy was not safe in this period; bands of run-away slaves (**fugitīvī**), such as those who made part of Spartacus's army, would be roaming the countryside.

Capua (an Etruscan foundation) was one of the largest and richest towns in Italy outside Rome.

Anxur was a small hill town some three miles from the terminus of the canal which ran from Forum Appiī. It was possible to travel overnight by barge on this stretch of the journey. Horace did so on his journey from Rome to Brundisium in 38 BC (*Satires* 1, 5). We take some details of the story from this passage.

Grammar

It is a defect of the Latin language that the same verb forms are used for the perfect and aorist tenses (Greek, French, English, etc. have two separate sets of verb forms). This failing in the Latin language will inevitably cause your pupils difficulty sometimes, and only the context will show which meaning is intended. Perfects always occur in a present context, and in narrative Latin they are mostly found in quoted speech. We hope that our examples make correct translation inevitable.

Exercise 14.5

p.115, l.2 **iam diēs aderat, cum . . . advēnērunt**: from now on we use **cum** + indicative whenever this gives correct Latin (**ubi** would be incorrect here).

Background

p.115 Regulus: Horace praised him in *Odes* 3, 5, 32ff.

p.116: in our map, we show not only Hannibal's route, but also the sites of his great victories over the Romans (Trebia, Trasimene and Cannae) and that of the battle of Metaurus where his brother Hasdrubal was defeated and lost his life (p.117, para. 4).

p.117, para. 2: the modern view is that

Hannibal was right to reject Maharbal's advice. He did not have the siege artillery to tackle the strong walls of Rome.
p.117 **dēlenda est Carthāgō**: the elder Cato had been sent on an embassy to Carthage in 153 BC where he saw how quickly the city had recovered from her defeat in the second Punic War. After this he repeatedly urged the Senate to strike again at Carthage before she once more became a menace to Rome; it is said that he ended every speech he made in the Senate with the words 'dēlenda est Carthāgō'. Eventually his policy prevailed and in 149 BC Rome declared war on a flimsy pretext.

There are various admirable books about Hannibal. Two of these are: Sir Gavin de Beer's *Hannibal* (Thames and Hudson); and Ernle Bradford's *Hannibal* (Macmillan).

Chapter 15

Captions: **cōnfectum, aedificia, satis** are new.

Grammar: the pluperfect tense

The pluperfect tense gives no trouble in form or meaning. In Latin it is used exactly as in English to indicate that an action is past time precedes some other past action.

Rōma

Quintus and Flaccus inevitably walk to the Forum, the centre of Rome geographically as well as politically, in their search for the school of Orbilius. The background section should be read before the story, so that the topography may make some sense and reference to the Vestal Virgins etc. may have some meaning. The **Cūria** had been burnt down by the time Quintus got to Rome; in January 52 following the murder of Clodius by Milo and his gang, the mob burnt his body in the **Cūria** (see background section in Chapter 17).

Lucius Orbilius Pupillus, a grammarian from Beneventum, started his school in Rome in 63 BC (Suetonius, *Gram.* 9). It had a high reputation, although Horace (*Epistles* 2, 1, 70) describes him as 'plāgōsum Orbilium' (Flogger Orbilius). We do not in fact know where his school was and have placed it rather vaguely.

The Subura was the poor, squalid, and crowded district of Rome lying between the Viminal and Esquiline hills. Augustus had a wall more than 100 feet high built in his forum to shut out the view of the Subura! For a description of life in the top of an **īnsula**, compare Juvenal, *Satires* 3, 193–203. Fire was a major hazard: 'If there is a panic at the bottom of the stairs, the furthest tenant will burn, who is only protected from the rain by tiles, where the soft doves lay their eggs.'

Exercise 15.5

★ p.123: a clothes shop in which two slaves are displaying material to a woman customer, in the Uffizi Gallery, Florence. Cushions are hanging from the rail.

toga praetexta (see background section of Chapter 18 for a brief description of the **toga**); the **toga praetexta** was worn by senior magistrates and freeborn boys until they assumed the plain white toga of manhood (**toga virīlis**). Refer your pupils to the illustration on p.11.

Background

The words of Camillus are taken from Livy 5, 54, those of Cicero from *Dē Rēpūblicā* 2, 11.

★ pp.125–6: the Temple of Vesta, a circular building with 20 Corinthian columns, was reconstituted to its present appearance (picture (a)) in 1930.

Chapter 16

Captions: **barbam, vultum, manū, versūs, medium, discipulum, commendō** can all be 'guessed' but NB **versūs** is accusative plural.

Lūdus Orbiliī

p.127 **Flaccus capsulam portābat**: the sons of wealthy parents were accompanied to school by a **paedagōgus** who carried their books and looked after them. Horace's father undertook this role (see *Satires* 1, 6, 81): 'My father himself was with me at all my teachers, an incorruptible guardian.'
p.127: Orbilius was a **grammaticus** and his school a grammar school, which provided the secondary stage of education; the **litterātor** provided the first. Grammar in the strict sense was taught, but the main part of the syllabus was the study of literary texts, both Latin and Greek. All educated Romans were by now bilingual (Quintilian even prefers that formal

education should begin in Greek rather than Latin); hence Orbilius is very shocked that Quintus knows no Greek.

The pupils had before them a papyrus scroll (**volūmen**), written with no word division or punctuation, entirely in capital letters, often peppered with mistakes; their first task was to correct (**ēmendāre**) and punctuate (**distinguere**). They would then read aloud (**recitāre**) with correct enunciation and appropriate expression. The teacher would give a commentary on literary, historical and philosophical points arising from the text. Pupils would often be required to learn parts by heart.

Considering the literature studied, mostly Greek (though Horace's own poetry, as he had feared (*Ep.* 1, 20, 17–18), became a school text by Juvenal's time (*Satires* 7, 225–7)), this was not such a narrow education as it sounds. The classical texts which survive today have done so largely because they were part of the Roman grammar school syllabus. This remained the staple secondary education until Rome fell and was revived from time to time thereafter, not so very much changed. Our own grammar schools, mostly founded after the dissolution of the monasteries, took over the system. The original statutes of these schools only included Latin and Greek in the prescribed syllabus; at some, e.g. Harrow School, boys were forbidden to speak English once they had left the first form. This diet was largely unchanged until the Grammar School Act of 1840.

p.128 **quōs**: relative pronoun as object: this will probably present no difficulty, since in translation the word order is the same in English. The relative is explained fully in Part II; it is best to skate over it at present.

p.129: the boy who befriends Quintus is Marcus Cicero, the great Cicero's son. In the next chapter he takes Quintus home and introduces him to his famous father. This is fiction; the young Cicero was born in the same year as Quintus (65 BC) but was educated under his father's supervision at home; he was rather idle, according to his father. He accompanied his father to Cilicia in 51 to 50 and was sent to university in Athens in 46. He was there at the same time as Quintus but studied (fitfully) at the Lyceum under Cratippus, whereas Quintus was at the Academy under Theomnestus. Both joined Brutus's army when he came to Athens recruiting from the students in September 44 BC. And so it is possible that they met some time in the period 44 to 42.

Exercise 16.4

Livius Andronicus (*c*.284–04 BC) was a Greek from Tarentum but was the founder of Latin literature. Lucius Livius brought him to Rome as a slave in 242 BC. There he was freed and took his master's name. He became a teacher of Greek literature, an actor, producer and writer of plays. He translated Homer's *Odyssey* into Latin and Greek tragedies. Here is the opening line of his *Odyssey*:

> Virum mihi, Camēna, insece versūtum

Written in the old accentual Latin metre (Saturnian), it is pretty uncouth stuff, but it is the first beginning of Latin literature.

Horace says that he studied Livius's poetry under Orbilius; in a passage in which he is complaining that critics praise the old poets simply because they are old and criticize contemporary poetry simply because it is new, he says:

> I'm not attacking the poetry of Livius or saying that it ought to be destroyed, the poetry which I remember Flogger Orbilius dictating to me when I was small . . .
> I am indignant that anything should be faulted not because it is thought to be clumsily or inelegantly written but because it was written recently – whereas the critics demand not pardon for the old poets but prizes of honour. (*Ep.* 2,1, 69ff.)

There is a description of Horace's personal appearance in Exercise 16.4, para. 2: **ego parvus sum et pinguis**. He describes himself (*Ep.* 1, 20, 24) as 'corporis exiguī' (of tiny stature) and (*Ep.* 1, 4, 15) as 'pinguem.' These derogatory self-descriptions are confirmed by a letter written to him by Augustus: 'I think you must be afraid of your books being bigger than yourself. You are lacking in height, not in girth; well, you can write on a pint pot, so that the circumference of your volume will be elephantine, like your stomach.'

Background

Further explanation of the political terms used here may be necessary: republic, senate,

senators, magistrates, consuls, tribunes, equestrians/knights, provinces. The word 'magistrate' especially needs clarification since a Roman magistrate was a state official with many (not just legal) functions. A province was, of course, one of the territories outside Italy governed by Rome.
p.133 **pānem et circēnsēs**: Juvenal 10, 81.

Chapter 17

Captions: **Campum** must be given and explained; **iuvenēs** can be 'guessed'.

This chapter introduces no new grammar and should present little difficulty.

Quīntus ad Campum Martium adit

p.136 **mihi nōmen est Marcus**: my name is Marcus (literally 'the name to me is Marcus'). This idiom is similar to the dative of possession: **est mihi equus** I have a horse. We do not discuss this usage in Part I and this is best treated as an isolated idiom at present.

Quintus's friend is gradually revealed as the son of the famous Cicero, to whom the background section of this chapter is devoted.
p.136 **Campus Martius** (see background section in Chapter 18): this large open space in the bend of the Tiber north west of the Capitol had been used in early times for the review and training of the army; hence its name. In Republican times it became the meeting place for the Comitia Centuriāta, the assembly which elected the major magistrates, and the place where the youth took exercise.
p.137: Marcus takes Quintus to visit his home on the Palatine hill. This was the most fashionable and expensive district of Rome, where Augustus later built his **palātium** (hence English 'palace'). Cicero's house had been destroyed by Clodius and his gang during Cicero's exile (see background section). It was rebuilt by decree of the Senate after Cicero's return in 57 BC despite Clodius's attempt to undo the work.

The fictitious meeting between Quintus and Cicero must be placed in late spring 51, quite soon after Quintus arrived in Rome; for Cicero left for Cilicia early in May, taking Marcus with him, and after his return in November 50 did not enter Rome, for fear of forfeiting the triumph which he was so keen to get.

When the boys enter his study, Cicero is dictating a letter to Atticus. Atticus and Cicero had been at university together in Athens and were life-long friends. Cicero had great respect for Atticus's judgement and relied on him for help and advice. He wrote to him continually on every topic under the sun – politics especially, but also on his family and financial affairs (Atticus was his banker), and on literary matters (Atticus was a keen critic and publisher). He sometimes wrote as many as three letters to Atticus in a day. These letters were collected and published by Atticus after Cicero's death in sixteen books.

Cicero's secretary was Tiro, a freedman who became a close and trusted friend. When Cicero was on his way back from Cilicia, Tiro fell ill and had to be left in Greece; Cicero wrote several letters to him at this time, which show deep concern and affection. Tiro collected and published Cicero's letters *ad Familiārēs* (to his friends) after Cicero's death (in sixteen books).

★ p.140: the **ātrium** of the Samnite House, Herculaneum (second century BC). Note the **impluvium**.

Exercise 17.6

Cicero's library: he was extremely proud of this, going to great lengths to obtain rare books and often appealing to Atticus to help him track them down or have a copy made. The young Marcus was a disappointment to his father. He was lazy at his studies, had no academic interests and drank too much.

Background

The words 'philosophy' and 'philosopher' may need explanation.

Cicero's sense of humour is commented on in various ancient writers, most notably Plutarch (*Comparison of Demosthenes and Cicero* 1). Here is a Ciceronian joke:

> In one of his speeches against Verres (2, 4, 132), Cicero talks about those who made a living from guiding tourists round Syracuse. After Verres had looted the place, he says, you would expect them to be out of a job. Not at all. Instead of showing the tourists the art treasures, they simply point to the places where they used to be before Verres removed them!

Chapter 18

Captions: **vōce** can be 'guessed' (**vocō** is known); **balnea** must be given in (2); **piscīnam**, **saliēbant**, **pīlīs** can be deduced from the pictures.

Grammar: uses of the ablative case

The uses of the ablative case are a major stumbling block. So far it has been possible to explain case usages very simply (though inadequately), but now we find that the ablative case has a bewilderingly wide range of uses, for which the traditional summary of meanings – 'by, with or from' – is quite insufficient. Your pupils will be able to master these only by experience and teachers will need to give help quite frequently and probably with more practice than we provide in the exercises.

Grammar: dative verbs

The datives used after these verbs are indirect objects. We can divide the verbs into two classes:

1 genuinely intransitive verbs, e.g. **hoc mihi placet** this is pleasing to me (**placeō** can never be used transitively); so also, **tibi resistō** I stand up to you;

2 verbs which can take a direct as well as an indirect object, e.g. **hoc tibi persuadeō** I persuade this to you, i.e. I persuade you of this (the verb is used transitively).

However, they are best learnt at this stage simply as verbs which are followed by the dative, though the dative should always be analysed as an indirect object (the principle that the direct object goes into the accusative must never be undermined).

Quīntus lūdō Orbiliī discēdit

p.148 the baths: see background section. Most Romans, unless they had private bath suites, would visit the baths every day, not only to bathe but also to meet their friends in what was virtually a large club. There were, of course, many sets of public baths in Rome and people would have their favourite set.
p.148 **nōnne tibi persuādēre possum mēcum venīre: persuādeō** + infinitive is legitimate classical Latin.

★ p.148: hot room in the Forum Baths, Pompeii. The fluted ceiling carried away the moisture from condensation; the floors and walls are hollow to allow the circulation of hot air.

p.149: Marcus drunk. He had a reputation for hard drinking; it was said, rather unkindly, that the only thing he would be remembered for was his ability to drink a gallon of wine at a go (Pliny, *Natural History* 28, 147).
p.149 **vēr aderat**: this would be spring 48 BC. In this chapter we have passed over world-shaking events. Julius Caesar crossed the Rubicon in January 49 BC and invaded Italy. There followed four years of civil war; the war was fought in the provinces – Spain, Greece (Pharsalus, 48 BC), Africa and finally Spain once more. We suppose that Quintus's education was unaffected by these events and indeed they may have seemed pretty remote to the ordinary man in Rome.
p.149 **dē Naeviō**: Naevius, born *c.*270, was the earliest native Roman poet. He was born in Rome, served in the first Punic War, and began to write at about the age of 35. He wrote comedies (both adapted from Greek New Comedies and comedies on Roman themes), tragedies based on Greek plays, and the first Roman historical epic, *Bellum Pūnicum*.

The following lines show the difficulty of this genre, which became very popular at Rome, though, fortunately perhaps, very little survives:

> trānsit Melītam
> Rōmānus exercitus, īnsulam integram ūrit,
> populātur, vāstat, rem hostium concinnat.

(The Roman army crosses to Malta, sets fire to the whole island, ravages it, lays it waste, settles the enemy's business.)

No wonder that Horace complains of contemporary critics:

> Isn't Naevius in their hands and embedded in their hearts as if he were quite recent? So revered is every ancient poem. *Ep.* 2,1,52–3

Nor is it surprising that Quintus's attention should have wandered from Naevius, as he made a first attempt at one of his most famous odes (4,7), of which A.E. Housman said, 'That I regard as the most beautiful poem in ancient literature.' **diffūgērunt**, of course, will not scan; the line is a dactylic hexameter.

Grammar: the declension of 'is' and 'ille'

Both can be used as adjectives as well as pronouns. They are not interchangeable, although we have had to give the same meanings for each of them; their usage is subtle and will only be grasped by meeting them in context. We hope we have used them correctly in our narrative. One of the commonest uses of **ille** is at the beginning of a new sentence to indicate a change of subject. One of the common uses of **is** is with the relative: **is quī** the man who . . .

Exercise 18.6

The rhetor, Heliodorus: on his journey to Brundisium (*Satires* 1,5) Horace was accompanied by Heliodorus: 'rhētor comes Hēliodōrus, Graecōrum longē doctissimus'. We have supposed that he kept a rhetorical school in Rome.

Rhetoric, which is not explained in the pupils' book, was the staple further education to which well-to-do boys proceeded on leaving school. Philosophy was, so to speak, an optional extra, which was taken by relatively few. Rhetoric was developed as a teachable skill by the sophists in the last quarter of the fifth century in Greece. It involved learning how to argue a case clearly, lucidly, and elegantly, and has been called the art of persuasion. The whole study was reduced to a complex system with many technical terms and rules. Students studied models, composed declamations on set subjects and took part in debates, often on imaginary law cases. Its practical importance in a world without newspapers, radio, or television can scarcely be exaggerated; the public speech was the only way of communicating with your fellow citizens in the mass and any Roman who hoped to make his mark had to strive to excel in this art. In Horace's day, debates divided into (1) abstract, general themes, and (2) particular themes related to a situation. You might find it a lively activity to discuss with your pupils in category (1) 'Should one marry?' and in category (2), with your pupils imagining themselves as the young Hannibal in Carthage, 'Should I cross the Alps in order to invade Italy?'

★ p.153: this bronze statue of an orator is an Etruscan work dating from the fourth or third century BC. The figure, known as the 'Arringatore' (the Italian word for 'orator'), wears a semicircular toga over a sleeveless tunic. The purple stripe is clearly visible. (Archaeological Museum, Florence)

Background

The quotation from Pliny at the top of p.154 is from *Letters* 1, 9.

Seneca (*Letter* 56) vividly describes the drawbacks of living near the baths. The noise, he says, was appalling. Vigorous body-builders groaned and panted, masseurs energetically slapped their victims, and the scorer shouted out the totals at ball games. There was the noise of quarrelling, of a thief being arrested (probably running off with someone's clothes), and of the splashing caused by those who insisted on hurtling themselves into the water. Additional hazards were those who liked singing in the baths, and the sellers of cakes, sausages, and other foods, advertising their wares each with his own noisy call.

★ p.154: Roman meals. The bread was found in the shop belonging to the baker Modestus in Pompeii. The fish, eggs, nuts, and lentils were found on a stove in the kitchen in the apartments of the priests of Isis: they formed the priests' last meal. (Behind the kitchen was found the skeleton of a man trapped on a narrow staircase, trying to escape from the rain of ashes; near him lay an iron mace with which he was trying to break out. The other priests did escape from the temple precinct but fell at various stages, overcome by the fumes; they are identified by the sacred objects they were trying to carry – a statue of Isis, a silver plaque, a silver urn showing the rites of Isis; their bursar died still clutching a sack of gold coins from their treasury.)

Chapter 19

Idūs Martiae

The year is 44 BC. We do not recommend an explanation of the cumbersome system of Roman dates at this stage.
p.158 **theātrum Pompēiī**: Pompey built this theatre in 55 BC; it was situated on the Southern edge of the Campus Martius, beside the Via Triumphalis, and was the first stone theatre in Rome. Since the Senate House had been burned down, the theatre was often used for

meetings of the senate.

The description of the assassination of Julius Caesar follows Suetonius (*Dīvus Iūlius* 82). His last words, as he saw Brutus running at him, were in Greek, according to Suetonius: 'You too, my son.' Rumour had it that Brutus was Caesar's illegitimate son, but the Greek word (*teknon*) is often used by an older man addressing a younger. The burning of Caesar's body in the forum: based on Suetonius (*Dīvus Iūlius* 84).

p.159 **in medium forum**: into the middle of the forum; compare **summus mōns** the top of the mountain. This idiom will probably require some comment.

Exercise 19.4

patrem . . . vocāvit: the object is separated from the verb which governs it by the relative clause. This may cause difficulty. This type of word order will be practised in Part II.

Background

A certain amount of geographical assistance may be necessary here (Gaul, the Rubicon, the Adriatic, Pharsalus).

We have highlighted Cicero in our treatment of the great political figures of the day. Teachers may wish to stress the stature of Julius Caesar. Pre-eminent as a general and statesman as well as a noble and cultivated man, he was one of the most impressive figures of the ancient world. 'Why, man,' exclaims Shakespeare's Cassius, 'he doth bestride the narrow world Like a Colossus.' Cassius was right.

★ p.164: the portrait of Antony is on a seal ring of red jasper dating from about 40 BC; the cameo of Octavian is in sardonyx, first century AD. It is from the Carlyle Collection.

Chapter 20

Grammar: the future tense

The endings of the 1st and 2nd conjugations differ strikingly from those of the 3rd and 4th. Your pupils will not find this easy to assimilate and may need more practice then we have provided.

Quīntus Rōmā discēdit

Heliodorus's outburst: Heliodorus expresses here, in appropriately rhetorical form, sentiments which occur repeatedly in the literature of this period, as the poets look back with grief and revulsion at the endless cycle of civil wars (e.g. Horace, *Odes* 1, 2; Virgil, *Georgics* 1, 489–514). It is worth listing these wars, which in the end destroyed the Republic:

1	90–89	The Social War (i.e. the war in which the Italian allies fought for Roman citizenship).
2	88–87	Marius versus Sulla and Optimates.
3	83–82	Sulla versus Cinna and Populares.
4	49–45	Caesar versus Pompey (Pharsalus 48) and senate.
5	43	Antony versus Octavian and senate (Mutina).
6	42	Antony and Octavian versus Brutus and Cassius (Philippi 42). Octavian versus sons of Pompey, etc.
7	32–31	Octavian versus Antony (Actium 31).

It has been said that there was not a family in Italy which did not suffer losses in lives or property or both; at the battle of Philippi alone there were over 24,000 dead.

p.168: Theomnestus was head of the Academy. The Academy was founded by Plato in 387 BC as a school for future statesmen ('philosopher kings'). Here he taught the philosophy of which the surviving dialogues are a 'popular' version. Although its reputation fluctuated, the Academy continued to function as an institute for research (especially in mathematics) and higher education until it was closed by Justinian in AD 529.

p.168: Ostia was the busy port of Rome, but at this time it had no adequate harbour (one was built by Claudius 100 years later), and ships moored at quays on the beach. Quintus had to take a small ship to Puteoli, the principal port on the East coast of Italy, and change there for Greece.

p.168 **advēneris**: the future perfect is introduced here for the first time without comment. The person is clear and the meaning should cause no difficulty; there are three more examples in the last paragraph of the narrative. The grammar and usage are explained after the narrative.

★ p.169: tomb painting from Ostia. The ship is named as the **Isis Giminiāna**. The captain, **Farnācēs**, stands on the **puppis**, holding the steering oar. The owner (?), **Arascantus**, supervises the loading of the cargo (marked **rēs**), which is clearly corn. Beside him stands a winged figure holding a wand, presumably Mercury, the god of gain and good luck.

Exercise 20.8

Quintus will return to Italy after studying at Athens and serving in Brutus's army at Philippi. Horace does not tell us what happened to his father and family.

> Philippi sent me off, humbled, with wings clipped, and robbed of my father's home and farm.
> *Ep.* 2,2, 49–51

Much land was confiscated from eighteen Italian towns, including Venusia, to provide for demobilized veterans after the civil war. We presume that Flaccus lost his farm (as Virgil's father may have done) in these confiscations and that Quintus's family disappeared without trace in the general chaos of these years.

Background

There will not be time to discuss all the aspects of the achievement of the Greeks and Romans, but teachers will probably wish to pursue two or three of the leads given in our material. Here, as elsewhere in the course, slides, film-strips, and video tapes would prove of great value.

★ p.174: this aqueduct brought water to the city of Nîmes in the South of France from the mountains twenty five miles away. Where it crosses the valley of the river Gard it is built on a triple arcade to a height of 180 feet above the river.

★ p.175: for more information about the Parthenon, see Part II, Chapter 2, Background, and Teachers' Book, p.34.

PART II

In the Teachers' Introduction to Part I, we explain the aims and scope of the course, the principles on which we have constructed it, and the way in which we intend it to be used. We recommend teachers who have not used Part I to read this before reading the Introduction to Part II.

THE LINGUISTIC CONTENT

Part I should have established firm foundations for the basic sentence patterns of Latin, as well as for the grammar of the verb (active voice), the declension of regular nouns and adjectives, adverbs and common prepositions and pronouns. Part II introduces the passive voice, participles and most types of subordinate clause in succession. By the end of Part II most of the grammar and syntax required for GCSE has been covered.

The gradient of difficulty remains constant and fairly shallow for the first eleven chapters. In the last nine chapters the gradient is rather steeper. Extracts from Latin authors, including some of Horace's poetry, are introduced from Chapter 12 onwards. These extracts are carefully chosen to fit the context of the narrative, and in trials they have occasioned less difficulty than we anticipated. By the end of Part II the transition to simple texts should not prove too traumatic.

Sequence of grammar and syntax

Chapter 1

Comparison of adjectives.
Relative clauses.

Chapter 2

Present participle.
Comparison of adverbs.

Chapter 3

Relative clauses continued.

Chapter 4

Perfect participle passive.
Ablative of instrument and agent.
Uses of the dative case.

Chapter 5

Subjunctive mood, present and imperfect.
Final clauses.

Chapter 6

Passive voice, perfect and pluperfect.
Final clauses (**nē**).

Chapter 7

Passive voice, present, future, imperfect.

Chapter 8

Deponent verbs.

Chapter 9

The ablative absolute.

Chapter 10

volō, nōlō, ferō.
Ablative of comparison.
Direct questions revised.

Chapter 11

Revision.

Chapter 12

Pluperfect subjunctive.
cum = when (in past time).
Indirect questions.

Chapter 13

Future participle; perfect subjunctive.
Indirect questions continued.

Chapter 14

Indirect statement.

Chapter 15

Indirect command.

Chapter 16

Revision.
negō, fīō, opus est.

Chapter 17

Consecutive clauses.
Numbers 200–1,000.

Chapter 18

Defective verbs (**coepī, meminī, ōdī**).
Verbs and adjectives + ablative.
cum = when (indicative in future time).

Chapter 19

Clauses after verbs of fearing.

Chapter 20

Revision.

THE EXERCISES

These follow a pattern similar to that of Part I. Sometimes grammatical explanation and exercises precede the narrative, sometimes they follow. The order is determined on purely empirical grounds, e.g. in Chapter 1 it seemed better to introduce the comparison of adjectives before the narrative, in which the comparative and superlative occur. On the other hand, the relative pronoun occurs frequently in the nominative case in Part I and so explanation and exercises have been placed after the narrative. In practice, it is no bad thing to give variety by spending part of a lesson on reading the narrative and part on doing exercises.

The last exercise of each chapter, as in Part I, consists of a passage of continuous Latin which develops or continues the narrative. Some of this is set for translation, while comprehension questions are asked on the rest.

These questions are of several kinds:

(a) straightforward questions on the sense, which admit of a precise answer;

(b) questions on the sense, which are more speculative and open-ended; the first of these occurs in Chapter 2, Exercise 5.5;

(c) questions calling for a personal response from the pupil, which do not admit of a precise answer, e.g. Chapter 5, Exercise 7.4;

(d) 'grammar questions', including English derivations, recognition of parts of speech, case usages, etc., and in the last chapters, turning direct into indirect speech, rephrasing the Latin, etc.; and finally, three short sentences for translation from English into Latin, the vocabulary for which is to be found in the passage just read.

In Chapter 16 and in some subsequent chapters, a passage of verse occurs in the final exercise. It is not intended that pupils should tackle this unseen. The verse passage should always be treated orally, the teacher helping and leading discussion, before questions, or a translation, are attempted in writing. The questions asked on the poems gradually become more sophisticated, but never employ the language of criticism. They will be discussed in this Introduction as they occur, but it is worth saying at this point that in framing them we have had in mind GCSE National Criteria 2.1.2: 'The aim in the linguistic area should be . . . to read, understand, appreciate and make a personal response to some of the literature in the original language.'

THE BACKGROUND MATERIAL

It will be possible for teachers to pursue background topics with pupils in rather greater depth in this second volume of the course. Further reading has been suggested where appropriate. The following books will prove especially valuable:

Joint Association of Classical Teachers: *The World of Athens*, Cambridge.
Bruno d'Agostino: *Monuments of Civilization – Greece*, Cassell.
H. H. Scullard: *From the Gracchi to Nero*, Methuen.
Ronald Syme: *The Roman Revolution*, Oxford.
U. E. Paoli: *Rome, Its People, Life and Customs*, Longman.
Peter Connolly: *Greece and Rome at War*, Macdonald.
John Boardman, Jasper Griffin, Oswyn Murray (edd.): *The Oxford History of the Classical World*, Oxford.

COMMENTARIES ON EACH CHAPTER

These commentaries deal with linguistic points which cause difficulty to pupils and points which are not adequately clarified by glosses. Teachers will find that, as in Part I, the narrative sometimes introduces points of Latinity which have not been explained; e.g. in Part II the relative with the subjunctive (not explained until Part III) occurs intermittently. We draw attention to such occurrences in our commentary but we do not consider it necessary or desirable to explain the grammar, provided that your pupils understand the sense; e.g. in Chapter 10, Exercise 9, Brutus says to Horace, **lēgātō mortuō tē ipsum dēsignāvī quī legiōnī imperēs**. Here **ut imperēs** would have been intolerable Latin, but we reckon that most pupils will not fail to see that this must mean 'I have appointed you yourself to command . . .'.

We comment on points of history, giving the sources for our narrative where we have followed ancient authors, filling in the historical background, and saying where we have resorted to fictions which are historically impossible.

Where appropriate we make some critical comments on the poems we introduce, especially where such comments are relevant to the questions we have asked. We hope that teachers will find time to discuss the poems with their class even where we do not ask specific questions on them.

★ Cover: see note on the illustration on p.180 of Part II.

★ Title page: this illustration from Trajan's column shows the Emperor Trajan standing on a **tribūnal** (a raised platform) with two of his staff. Legionaries and auxiliaries, standing at ease, listen attentively. Note the Roman standards.

Trajan's column, one of the most famous monuments at Rome, was dedicated in 113 AD to commemorate the Emperor's conquest of the Dacians (the inhabitants of modern Rumania). 100 Roman feet (29.78 metres) in height, it is carved with a continuous spiral relief illustrating the Dacian campaigns (101–2, 105–6 AD). It is an unrivalled source of information about Roman warfare. The statue of Trajan on top of the column was replaced in 1588 by one of St Peter.

Chapter 1

Part I ends with Quintus's father Flaccus seeing him off at Ostia, the port of Rome. Quintus will have to change ships at Puteoli, the principal port on the west coast of Italy.

In Graeciam

p.10 **Acadēmiā**: founded by Plato in 385 BC, the Academy had a continuous existence until it was closed by the emperor Justinian in AD 529. After Plato's death, it was one of the two leading philosophical schools at Athens; the other was the Lycēum, founded by Plato's pupil Aristotle. When Horace went there, the Head of the Academy was Theomnestus.

p.10 **Pūblius**: i.e. Publius Vergilius Maro. Virgil – this is the conventional English spelling – was twenty-five in spring 44 BC (he was born on 15 October 70 BC). After his education in Cremona, Milan and Rome, where he studied rhetoric, he left Rome for Campania, to study at the Garden, the Epicurean school directed by Siro. He stayed there until Siro's death in 42 BC. If we wish to bring this imaginary meeting between Virgil and Horace into correct historical perspective, Virgil must be thought of as returning to Siro's school rather than starting there.

p.10 **Cūmae . . . Mīsēnum**: these sentences look back to Part I, Chapter 10.

Puteolī: now Pozzuoli, near Naples. This was the main port on the west coast of Italy until Claudius built the great harbour at Ostia. It was also a flourishing resort.

★ p.11: this shows the bar counter in the **caupōna** on the Via dell'Abbondanza at Pompeii. The bar counter is L shaped; it contains large jars for keeping food hot. Wine jars rest against the wall addressed 'to Euxinus the innkeeper, near the amphitheatre, Pompeii'. Behind the bar are three other rooms, a store room and a lavatory; on the right is a courtyard for eating and drinking in the open air.

★ p.12: the Straits of Messana between Sicily and the toe of Italy are 3 miles wide at their narrowest point.

Grammar: the relative pronoun

This has frequently been used in Part I in the nominative case but its use has not yet been explained. In this chapter we still use the relative only in the nominative case. Pupils will find difficulty not in the relative clause, but in the word order exemplified by Exercise 1.4.7, i.e. subject-object-relative clause-main verb.

To prepare for this you might vary the order of the second example (**nōnne templa quae in forō stant vīdistī?**).

Exercise 1.7

p.15, l.14 **duōs viātōrēs undae in mare rapuerant**: word order: object, subject, verb. This is liable to cause a stumble. Pupils must become accustomed to variations in word order if they are ever to read original texts.

From Agamemnon to Alexander

This background section is radically different from any of the others. It is intended primarily for reference, and its use is to provide historical perspective for the many references to Greek history which occur in the next few chapters, and to provide some idea of the background of the Hellenism which the Romans inherited. We make no apology for devoting so much space and time to Greece in a Latin course. Latin literature makes no sense in isolation from Greece, as the Romans well knew. See Horace, *Epistles* 2, 1, 156–7:

> Graecia capta ferum victōrēm cēpit et artēs
> intulit agrestī Latiō.

(captured Greece took its savage victor captive and brought the arts into rustic Latium.)

★ p.16: this gives an impression of what the megaron (or great hall) of a Mycenaean palace may have been like. The hearth in the centre was the focal point of the palace. Facing it, in the centre of a wall, was the king's throne. Floors, walls and ceilings were ablaze with colour. (See J. Bolton: *Ancient Crete and Mycenae*, Longman, pp.70–1.)

★ p.18: frieze from Persepolis. It dates from the early fifth century BC and is in the Archaeological Museum at Teheran.

★ p.19: this famous bust of Pericles is in the British Museum. He is thought to have insisted on being portrayed in a helmet in order to disguise his abnormally large cranium.

★ p.20: in 1977 the first unplundered Macedonian tomb came to light at Vergina in North Greece. This miniature ivory head, showing a mature bearded man, is probably King Philip. It is just over three centimeters high. (Archaeological Museum, Thessalonike.)

★ p.21: this mosaic from Pompeii, a Roman copy of a Greek painting, shows Alexander, his eyes blazing as he spurs his horse forward at the Battle of Issus (333 BC). It was in this battle that he broke the might of the Persians.

Section 9 Hellenistic: this word may need explanation. It refers to the era when Greek civilization spread and became dominant within the boundaries established by the conquests of Alexander.

If teachers should wish to amplify our skeletal summary in some areas, a good account of Greek history up to the death of Alexander can be found in Joint Association of Classical Teachers: *The World of Athens*, Cambridge, pp.1–61.

Chapter 2

Grammar: the present participle

The present participle has frequently been used in Part I in the nominative case, but its occurrences have always been glossed. It gives little trouble when used agreeing with nouns in the nominative and accusative, but when it agrees with a genitive or dative it cannot often be translated literally, e.g. cartoon caption 2: **nauta Quīntō rogantī . . . respondet . . .** To Quintus asking . . . the sailor replied . . . . We have to say in English, 'When Quintus asked . . .'. There is another example of this in Exercise 2.5 which may give trouble. The same problem will occur in some sentences with perfect participles passive and the sooner pupils achieve a flexible approach to the translation of participles, the better.

Athēnīs

p.25, para. 3 **Parthenōna**: a Greek accusative form; best to gloss this over, unless questions are asked.

★ p.25: to the right of the Doric columns of the Temple of Hephaestus are the Parthenon, the Propylaea and the Temple of Athene Nike.

p.26: Quintus's letter home. This is the first example of a Latin letter and it might be useful to discuss the differences in format between English and Latin letters. Further examples occur in subsequent chapters in which some use is made of Cicero's letters.

p.27, para. 2 **rīdentium . . . loquentium**: the first examples of the present participle in the genitive. These can be translated by English participles ('a group of young men laughing . . .') or by English relative clauses ('who were laughing . . .'); pupils might be encouraged to offer these alternatives.
Marcus Cicerō: in Part I we make Horace meet Marcus, the great Cicero's only son, at the school of Orbilius. This is fiction; Cicero educated Marcus at home. Born the same year as Horace, he was studying in Athens at the same time, at the Lyceum under Cratippus. Later both served in Brutus's army. It is quite possible that they met during this period.
p.27, last para. **scholam**: **schola** can mean (1) a lecture (2) a school, especially a philosophical school. We use the word in both senses according to the context, which pupils may find confusing.

Exercise 2.5

Some discussion and development of the subject matter of this exercise would come well, preferably with illustrations.
portās magnificās: the Propylaea formed part of Pericles's great scheme for rebuilding the Acropolis monuments destroyed by the Persians: Propylaea (completed 432 BC); Parthenon (432 BC); Temple of Athene Nike (424 BC); Erectheum (after 395 BC).
p.28, last para. **interiōra templī obscūra erant**: the interior of all Greek temples was dark; there were no windows. The great statue of Athene Parthenon by Pheidias stood 11.5 metres high. It was probably made of wood and then covered with gold plates; the ivory portions (face, hands, etc.) must have been glued or tenoned into place (see R. J. Hopper: *The Acropolis*, Weidenfeld and Nicolson). The statue vanished in late Roman times but some models of it survive, which give a pretty inadequate idea of what it must have looked like. It certainly made a great impression on the Greeks and Romans who have left descriptions of it.
p.29 **fābulās clārās**: you will hardly have time for a major digression on Greek drama, but it would be worth remarking that it was in this theatre that the first tragedies and comedies in European literature were put on, the masterpieces of Aeschylus, Sophocles, Euripides and Aristophanes, which are still performed today.

★ p.29: the remains of the Theatre of Dionysus date from the reconstruction in stone – in the great age of Athenian drama the seats were probably made of wood – by Lycurgus in 342–326 BC. Its capacity is unlikely to have exceeded 17,000.

The theatre in the foreground is the Odeon of Herodes Atticus, built in the second century AD. The seating was entirely restored in Pentelic marble in 1950–61. This theatre can hold 5,000–6,000, and it is still used for dramatic, operatic and orchestral performances during the Athens Festival.

p.30, Question 5: the first comprehension question which requires a little more than a surface understanding of the sense. Quintus was impressed by the pre-eminence of the Athenians in the arts of architecture and sculpture and, by implication, literature. So was Virgil. See *Aeneid* 6, 847ff.:

> excūdent aliī spīrantia mollius aera
> (crēdō equidem), vīvōs dūcent dē marmore vultūs . . .
> tū regere imperiō populōs, Rōmāne, mementō . . .

(others, I believe, will beat out bronzes so that they breathe in softer lines, and will make living faces from the marble. You, Roman, remember to rule the peoples of the world beneath your sway . . .).
p.30, Question 6 ('In what case and why?') answers, e.g.: **manū**, ablative: in her hand. **templō**, ablative; governed by preposition **in. nāvēs**, accusative; object of **vidēre. poētae**, nominative; subject of **indūxerant**. The names given to case usages need not, in our view, be learnt.

Background

Supplementary material on the topography of Athens can be found in the JACT *World of Athens,* pp.78–88. We hope you may be able to supplement our illustrations with slides, etc. (and, if possible, a visit to the British Museum to see the Elgin Marbles).
para. 1 the Long Walls: if you take the train from Piraeus to Athens, for most of the way you follow the course of the southern Long Wall.
para. 2 Kerameikos: here Pericles, the great Athenian statesman mentioned in the

background section of Chapter 1 (Sections 5 and 6), delivered his celebrated funeral oration over the Athenians who had died in the first year of the war with the Spartans (430 BC: Thucydides 2, 35–46). The head and spear of the towering statue of Athene mentioned in the final paragraph became visible to sailors as they rounded Cape Sounion.

★ p.31: the Temple of Hephaestus is the best preserved of all Doric temples in mainland Greece. It overlooks the agora.

★ p.31: the Parthenon was erected between 447–32 BC. The famous frieze portrays the procession of celebrants at the Great Panathenaic Festival. Our photograph shows some animated riders. It is part of the West frieze, which remains *in situ*. Most of the rest of the frieze is now in the British Museum.

★ p.32: the delicate Ionic columns of the Temple of Athene Nike deserve particular attention.

Chapter 3

The relative pronoun in all cases

The main difficulty will be in the word order rather than in the relative clause itself, e.g. cartoon caption 1; here you will have to 'give' **tabellārius** (=postman). Ask what he is doing; the meaning will then be clear. Where difficulties arise, they can be solved by bracketing off the relative clause and analysing; this is required in Exercise 3.1.

Acadēmia

p.36 **ubi** = where? (this often causes a hiccough).
p.36, para. 3: Theomnestus's lecture 'quid est summum bonum?'
Philosophy at this time meant primarily what we call moral philosophy and theory of knowledge. Horace, in *Epistles* 2, 2, 41ff., describes this period of his life as follows:

> It was my good fortune to be brought up in Rome and to be taught what harm Achilles in his anger did to the Greeks (i.e. he studied the *Iliad*, literature). Good Athens gave me a little more culture, so that I was willing to distinguish the straight from the crooked (i.e. moral philosophy) and to seek for truth in the woods of the Academy (i.e. theory of knowledge). But harsh times moved me from the place I loved, and the tide of civil war carried me, a novice, against the arms of Augustus Caesar, for which my muscles were no match.

The passage breathes a strong affection for his old university.
p.36, last para.: Marcus was not a good student. He was very extravagant, idle and too fond of the bottle.
p.37, last para. **Tīrōnī**: Tiro was Cicero's freedman, confidential secretary and friend. How much Cicero depended on him is shown by the eleven letters he wrote to Tiro when the latter fell ill in Greece on their way back from Cilicia (November 50 BC) and had to be left behind (see Chapter 15, Exercise 5).

The letter is a simplified version of Cicero, *ad Fam*. 16, 21, i.e. a letter which **Cicerō fīlius** actually wrote to Tiro from Athens in August or early September 44. Cicero had been worried by the bad reports he was getting of his son's extravagance and idleness – his Greek teacher was said to have led him astray! A little later he actually set out for Athens to see whether Marcus was behaving himself (see Chapter 5).
p.38, para. 3 **cupiō illam rem tibi fēlīciter ēvenīre** I hope this turns out well for you. Unexpected Latinity, but it is what Marcus wrote.
p.38, para. 4: it may be necessary to make the point of this paragraph more explicit. (If Tiro persuades Marcus's father to send him more money, Marcus will be in a position to help Tiro pay for his farm.) It would be worth while to discuss the relationship between Marcus and Tiro which this letter suggests. Despite the smoke-screen Marcus is throwing up, his affection and respect for his father's secretary is plain and there is a nice element of teasing humour in the picture of Tiro becoming **rūsticus Rōmānus**. If you do this, you will have to make it clear that this is an actual letter from Marcus to Tiro, in which Quintus has no hand.

★ p.40: this mosaic panel representing the Academy was found near Pompeii. It belongs to the late second or first century BC. The figure pointing to the globe is Plato, who is leading a discussion on astronomy.

Background

The Stoic school was founded by Zeno of Cyprus (335–263 BC) who came to Athens early in life and taught there in the Painted Porch (Stoa), from which the philosophy derived its name. It had a greater effect on Roman life than any other philosophy, lasting from its introduction to Rome in the second century BC until the third century AD. It proved a humanizing influence in the treatment of slaves, whom Stoics believed to be equal with other men because all alike are the sons of God. The Stoic emperor Antoninus (86–161 AD) wrote: 'My city and my country, so far as I am Antoninus, is Rome, but so far as I am a man, it is the world.' 'Men exist for the sake of one another.'

The Epicurean philosophy tried to rid men of whatever stood in the way of peace of mind. The fear of death, for example, could be overcome by the belief that the soul dies with the body and there is no danger of survival. In his great poem *Dē Rērum Nātūrā*, Lucretius (94–55 BC) gives twenty-eight proofs that the soul is mortal. In the passage quoted on p.42 and in the subsequent lines, he celebrates the Epicurean's distance from the madding crowd's ignoble strife:

> But nothing is more sweet than to dwell in the serene temples built on the teaching of the wise, from where you can look down on others and see them wandering this way and that and going astray while they seek the path of life, striving with each other in their ability and their claims to noble birth, struggling night and day by supreme efforts to rise to the height of power and gain dominion over the world. *DRN*, 2, 7–13.

For further reading on Stoics and Epicureans, see Bertrand Russell's *History of Western Philosophy*, Unwin, pp.249–76. A good summary of Lucretius's Epicurean faith is given on pp.1–5 of E. J. Kenney's edition of Lucretius, *Dē Rērum Nātūrā* 3, Cambridge.
p.41, para. 3 virtuous: the best translation of the key Stoic word **virtūs** is perhaps 'manliness'. This allows it to include **cōnstantia** as well as **fortitūdō.**

Chapter 4

Perfect participles passive

In Part I a few perfect participles passive were introduced as ordinary adjectives, e.g. **parātus** ready, **territus** terrified, which should help pupils at this point.
NB The passive voice has not been mentioned yet and the difference between active and passive is not discussed explicitly until Chapter 6. This may cause problems but we suggest that the question is not raised unless pupils ask for an explanation.
p.44: ablative of instrument and ablative of agent. This gives no difficulty in reading Latin, though it does in translating English into Latin.

Exercise 4.2

The analysis of participial phrases can either be done as in the example or by joining the noun and participle which agree by a line, e.g.
puerī ā-magistrō territī . . .

Quīntus iter facit

Educated Romans regularly visited the famous sites of Greece and Asia Minor (cf. Catullus 4 – his return journey from Bithynia in his yacht – and 46: **ad clārās Asiae volēmus urbēs**). Pausanias (fl. 150 AD) wrote a guide book for the benefit of tourists with detailed descriptions of the sites.
p.46, para. 2 Xerxes's throne: see Aeschylus, *Persae* 465–7: 'And Xerxes, seeing the depth of his disaster, groaned aloud; for he had a seat from which he could clearly see his whole navy, a high hillock near the sea shore.'

★ p.46: the massive Doric columns of the sixth century Temple of Apollo tower beneath the citadel to the South.

p.46, para. 4: Servius's letter to Cicero (*ad Fam.* 4, 5). Servius Sulpicius Rufus, an eminent jurist (cos. 51) and old friend of Cicero, wrote this letter from Athens in March 45 to condole upon the death of Cicero's beloved daughter Tullia. The passage quoted offers what seems to us cold comfort; he continues. 'Are we poor mortal creatures indignant if one of us dies or is killed, when in one place the corpses of so many towns lie flung down? Come on, Servius, pull yourself together and remember that you were born a mortal.'

★ p.47: in this Attic red-figure vase, now in Boston, Agamemnon is enmeshed in the net flung over him by his murderers.

Grammar: uses of the dative case

3 The dative of interest (or 'the person concerned') has occurred in earlier chapters but has been glossed. It has a wide range of uses which are gradually introduced.

Exercise 4.5

The Olympic games were first held in 776 BC and thereafter every four years without break until 393 AD. They opened with a day of sacrifice and festivity; on the second day were held the chariot races and horse races and pentathlon; on the third day the boys' contests took place and on the fourth the men's foot-races, jumping, wrestling, boxing and the pancration (a form of all-in wrestling); on the fifth day there were sacrifices and a banquet. Apart from the athletic contests the Games were the occasion of a great Panhellenic gathering. By Horace's time, their prestige had sunk and some races, e.g. the chariot races, had been discontinued. Nero revived them in splendour and won every race.

★ p.49 the model: the Temple of Zeus stands at the centre; to its right is the older temple of Hera. The small buildings to the right are treasuries dedicated by Greek states. To the left are administrative buildings, and behind them a big guesthouse. Back centre is the workshop of the sculptor Pheidias, and to the right of that is the gymnasium. In the right foreground is the tunnel leading to the stadium.

Background

p. 50: for the Corinthian and the Saronic gulfs, refer your pupils to the map on p. 45. The Corinthian Gulf is to the west of Corinth, the Saronic Gulf to the east.
p. 51: the Byron stanzas are from *Childe Harold's Pilgrimage* (Canto 4, 44–6); second stanza, ll.3–4: the Turkish shanty towns make the ruins of a once great and civilizing city both more sad and more precious.

★ p.52: the gateposts of the Lion gate are 10½ feet high; on top of these is a massive lintel, 15 feet long, 6½ feet thick and 3¼ feet high in the centre. A huge triangular slab of grey limestone (12 feet wide at the base, 10 feet high and 2 feet thick) shows a pillar supported by two lionesses which rest their feet on the altars that are the base of the column. The heads are missing.
p.53, para. 1 countless statues: in the time of Pliny the Elder there were as many as 3,000 statues in the enclosure.
p.53, para. 2: the truce was in fact broken on two occasions. In 420 BC the Lacedaemonians were banned from the festival for truce-breaking (Thucydides 5, 49), and in 364 BC a battle was fought in the enclosure before the crowd which had come to watch the games.

Chapter 5

The subjunctive mood

The uses of the subjunctive in main clauses (jussive, optative, potential, deliberative) are not explained until Part III. We have given the conventional meanings of the subjunctive – present 'I may', imperfect 'I might'. These meanings work for English final clauses.

Clauses of purpose (= final clauses)

We do not give the full rules for sequence of tenses. Before pupils attempt Exercise 5.4 they should be warned that if the main verb is present, future or imperative, the subjunctive in the **ut** clause is present; if the main verb is past, i.e. imperfect, aorist (which we have called 'perfect'), or pluperfect, the subjunctive is imperfect. We have avoided giving sentences where the main verb is a true perfect, e.g. We have come that we may see you **vēnimus ut tē videāmus**. In fact the sequence of tenses in Latin is the same as that in English where the conjunction 'that' is used.

Brūtus Athēnās advenit

p. 57, para. 2: Decimus appears in Part I as a schoolfellow of Quintus in Venusia (Chapter 5). He was then a bully and notoriously thick; he has evidently changed for the better.
p.57, para. 4 **Octāviānus**: when Octavius took his adoptive father's name (C. Iulius Caesar), his own name automatically changed to Octavianus. In historical fact, while he was referred to as Octavianus in 44 and 43 BC, he would have normally been known as Caesar until he assumed the name Augustus in 27 BC.

We have called him Octavian up to that time in order to avoid confusion with Julius Caesar.
p.58, para. 3: Marcus's letter from his father. Cicero was so worried about Marcus that he decided to visit him in Athens. He sailed from Pompeii on 17 July but was driven back from Syracuse by contrary winds. News from Rome, in particular the news that the Senate was to meet on 1 September, convinced him that he should return to Rome. On 2 September, he delivered the *First Philippic*, a strong attack on Antony's recent policy.
p.58, para. 6: Brutus's visit to Athens; Brutus left Italy at the end of August after an open quarrel with Antony. He arrived in Athens in September; he attended the lectures of both Theomnestus and Cratippus, hobnobbed with the students, and won over many to his cause, while at the same time his agents were winning over to his side Macedonia and its army.
p.58, last para.: Harmodius and Aristogeiton; according to popular tradition, these two young nobles liberated Athens from the tyrant Hippias, son of Pisistratus in 514 BC. They were honoured as the Liberators; their statues were set up in the Agora. In fact, the brothers killed Hippias's younger brother Hipparchus, and Hippias was driven out of Athens three years later (Thucydides 1, 20).

Brutus's arrival was greeted with enthusiasm by the Athenians, who decreed that statues of him and Cassius should be erected beside those of Harmodius and Aristogeiton.

★ p.59: the original statues of the tyrant slayers were carried off by Xerxes in 480 BC. Our illustration shows good copies of the statues which replaced them in 477. (They are in the Museo Nazionale at Naples.)

Grammar: perfect participle passive (contd.)

Pupils are likely to find difficulty in translating sentences of this form and will need practise.

Exercise 5.6

9 and 10: **eī quī** . . . those who; **ea quae** . . . the things which; **is** and **quī** are often used as correlatives, an idiom which may cause a stumble.

Exercise 5.7

p.61, last para.: Brutus thought very highly of Marcus Cicero; according to Plutarch, he said 'he could not help admiring a young man of such spirit and such a hater of tyranny'.
Question 4 asks for a personal response; such questions are not really suitable for written answers and might be better treated orally.

Background

The *Philippics*: Cicero himself called his attacks on Antony by this name, thus bringing to mind the great – and equally fruitless – speeches of the Athenian orator Demosthenes warning his fellow citizens against the expansionist aims of Philip of Macedon. Cicero's *First Philippic* (2 September) was restrained and moderate, but Antony's response was a violent attack on Cicero (19 September). This injected the vitriol into Cicero's subsequent *Philippics*. Here Cicero attacks Antony intemperately: 'Down those jaws of yours, down those lungs, with that physical stamina worthy of a gladiator, you sank so much wine at Hippias's wedding that you had to vomit in full view of the people of Rome the morning after. It was a revolting sight; it was revolting even to hear about it. If this had happened at a dinner party when you drain your notoriously huge goblets, it would still have seemed disgusting. But in an assembly of the Roman people, you, a man holding public office, a Master of the Horse – from whom even a belch would have been deplorable –, you, I say, filled your lap and the whole platform as you threw up chunks of food stinking of wine.'
(*Phil.* 2, 63).
p.62, question: Cicero means, of course, that he would have made sure that Mark Antony was killed as well as Caesar.

Chapter 6

Passive voice

The concept of the passive voice is not as obvious to young pupils as it is to adults. It might pay to give some practice in turning English sentences from active to passive form before tackling the Latin, e.g. Tom kicked the ball hard = the ball was kicked hard by Tom. Julia quickly ate the sweets = the sweets were quickly eaten by Julia.

Tenses

Pupils will probably grasp the meaning of the captions, since they have already met perfect participles passive and ablative of the agent. But they may well be confused by the tenses and will probably translate **audītus est** 'is heard'. (cf. **cēna parāta est** Dinner is ready).

The truth is that the so-called perfect and pluperfect tenses passive have two distinct meanings depending on whether the participle is used in a true perfect or an aoristic sense, e.g. **cēna parāta est** Dinner is ready (participle is used in a true perfect sense).
cēna ā mātre celeriter parāta est Dinner was quickly prepared by mother (participle is used in an aoristic sense). This is illustrated by the second caption:
iuvenēs, quī valdē commōtī erant, ā Brūtō salūtātī sunt The young men, who were deeply moved, were greeted by Brutus.
(**commōtī** appears to be a true perfect participle passive, **salūtātī** aoristic.)
But in practice this ambiguity does not often cause difficulty and, if you do the first examples of Exercise 6.1 orally, pupils will probably be able to cope.
p.65, para. 1 **Cicerō ōrātiōnem** . . .: Cicero attacked Antony on 2 September (the *First Philippic*). See chronological table on p.39.
p.65, para. 2 **Pompēius**: the Pompeius referred to is the friend whose homecoming Horace celebrated in *Odes* 2, 7 (see Chapter 13). Horace and he served in Brutus's army together. We know nothing about him except what we learn from this ode.
p.66, para. 2 **duās rūpēs ingentēs**: the Phaedriades (the Shining Ones) dominate the site of Delphi (see illustration).
p.67, para. 1 **aedem quae ab Athēniēnsibus exstructa erat**: the Treasury of the Athenians, now beautifully restored, was in fact built to commemorate the battle of Marathon soon after 490 BC. A little higher up the Sacred Way is the Stoa dedicated by the Athenians after the end of the Persian Wars.
p.67, paras. 3 and 4: the procedure for consulting the Delphic oracle is not known in detail: 'None of our ancient sources gives a straightforward account of what happened at Delphi when a consultation took place.' (Parke: *Greek Oracles*, p.80). Our account follows Hoyle: *Delphi*, Chapter 3.
p.67, last para.: on the walls of the temple of Apollo were inscribed the maxims: 'Know thyself' and 'Nothing in excess'. We have made Horace know at this moment that he will be a poet. The maxim is usually taken to mean 'Know your mortal nature'.

Locative case

Locatives of third declension nouns singular; early Latin used the ending **-ī** (instead of **-e**) and this remained common with **rūrī** in the country. We have omitted this complication at present.

Exercise 6.7

p.69, para 1 **fōns Castalius**: the Pythia and those who had come to consult the oracle purified themselves with water from this spring. Later it became known as the source of poetic inspiration.

Background

p.71: Apollo's declaration is from the *Homeric Hymn to Apollo*, 247–53.
p.71, para. 5: there is no reliable evidence for the traditional belief that the Pythian priestess worked herself up into a state of ecstasy while munching a laurel leaf and then babbled confusing prophecies as she sat on a tripod over a chasm which belched fumes (see ed. P. E. Easterling and J. V. Muir: *Greek Religion and Society*, Cambridge, Chapter 6, and JACT: *World of Athens*, pp.98–100).

★ p.72: Greek states built treasuries at Delphi to house their offerings to Apollo. At the same time they aimed to make propaganda points, especially through the beauty and expense of their buildings. The Treasury of the Athenians was built soon after 490 BC with a tithe of the spoils of Marathon. It was reconstructed by the French in 1904–6.

Chapter 7

Passive voice: present, future, imperfect tenses

The task of learning these is made much easier if your pupils first study the table of active and passive person endings. The tenses and persons will then be immediately recognizable in Latin.

In studying these tenses, one difficult point must be stressed: in the third conjugation

regeris (short **e**) is 2nd person singular *present*, **regēris** (long **ē**) is 2nd person singular *future*.

The passive endings in our table apply to the present and imperfect subjunctives passive as well as the indicatives. If your pupils find any difficulties with these subjunctives, they are of course given in the Summary of Grammar (p.222).

Bellum Cīvīle

The events between the assassination of Julius Caesar and the death of Cicero are so complex that a chronological table may be useful:

44 BC	
15 March	Assassination of Julius Caesar.
April	Octavius returns to Italy; adopted by Caesar's will, he becomes C. Julius Caesar Octavianus. Antony (consul) rebuffs Octavian and strengthens his position. August: final rupture between Antony and Brutus and Cassius; Brutus goes to Athens and Macedonia, Cassius to Syria.
2 September	Cicero attacks Antony in Senate (First Philippic).
October	Octavian raises troops from Caesar's veterans and marches on Rome; armed conflict with Antony narrowly averted.
December	Antony leads forces to Cisalpine Gaul; when D. Brutus refuses to hand over Mutina, Antony invests the town.
20 December	Cicero delivers the Third Philippic (to the senate) and the Fourth (to the people). He comes forward as leader of the Republic against Antony.
43 BC	
January	The Senate sends an embassy to Antony, ordering him to evacuate Italy. Octavian is made a senator and given propraetorian imperium.
February	Senate declares state of war.
March	Both consuls (Hirtius and Pansa) and Octavian march against Antony.
April	Mutina campaign. Antony defeated; he withdraws to Transalpine Gaul. Both consuls killed in battle.
May	Antony wins support of Lepidus (governor of Narbonese Gaul) and regains control of Cisalpine Gaul.
August	Octavian marches on Rome and demands consulship; elected 19 August.
October	Octavian marches North, ostensibly to oppose Antony. He meets Antony and Lepidus at Bononia; Second Triumvirate formed.
November	The Triumviri in Rome; 27 November, Lex Titia makes them **IIIvirī reīpūblicae cōnstituendae**. Proscriptions.
7 December	Cicero killed.

p.75: Marcus's letter to Quintus is fictional.
p.76, para. 2 **imperium**: this is a technical term. **imperium** was the supreme administrative power, involving command in war. Originally only consuls and praetors had **imperium**; later proconsuls and propraetors held it and it was occasionally conferred on other individuals by a decree of the senate. That is what happened in the case of Octavian, who held no magistracy.
p.76, para. 3 **iam imperium eī adimere in animō habēbant: eī** from him (dative of person concerned or of disadvantage). Cicero is reported to have said 'laudandum adulēscentem, ornandum, tollendum' (*ad Fam.* 11, 20, 1) (the youth should be praised, honoured, and then removed). This remark is said to have reached Octavian's ears.
p.76, last para.: the succession of events was in

fact a little more complex than this. In July Octavian sent a request to the Senate asking for a dispensation allowing him to stand for the consulship, which was refused. He then marched on Rome at the head of eight legions. He entered Rome unopposed and was elected consul by the people on 19 August at the age of nineteen. His election was contrary to the law, by which magistracies had to be held in succession and at minimum ages.

Exercise 7.5

This account is based on Livy, frag. 120.

Background

p.80: the death of Cicero. It may seem strange that Octavian agreed to the proscription of Cicero, a man who had given him so much support. Octavian was implacably determined to take revenge on all the conspirators who had killed Julius Caesar. While Cicero had not been a party to the conspiracy, he had rejoiced, even gloated over Caesar's assassination, and this, in conjunction with Cicero's rash pronouncement that Octavian was a young man to be made use of and then cast aside, may well have made Octavian ill-disposed to intercede on his behalf.

Chapter 8

Deponent verbs

Deponent verbs appear rather close on the heels of the first introduction of the passive voice. The list of deponent verbs will have to be carefully learnt if confusion is to be avoided.

Quīntus mīlitat

We do not know precisely when Horace decided to join Brutus. In delaying his decision until after Cicero's death, three months after Brutus's visit to Athens, we have perhaps stretched things a bit.
p.83, para. 2 **Brūtus exercitusque iam in Asiā sunt**: late in 43, after consolidating his position in Macedonia, Brutus crossed to Asia to meet Cassius, who had been in Syria. They decided to raise money in Asia, and Brutus forced the cities of Lydia to contribute. When Xanthus, a city in Lycia, refused to cooperate, he besieged it and took it by storm. Horace was certainly with Brutus in Asia (see *Satires* 1, 7).
p.83, para. 4 **deōs ōrō ut . . . redeās**: an indirect command, a construction not yet explained, but the meaning should be clear (e.g. **veniō ut tē iuvem** I am coming to help you, **tē rogō ut mē iuvēs** I ask you to help me.)
p.83, penultimate para. **Dēlos**: Delos had become the centre of the slave trade in the Eastern Mediterranean, where thousands of slaves changed hands each day. By the time Quintus visited it, this trade was declining and by the end of the century the island was virtually uninhabited.
p.84, para. 1 **palma alta**: Leto was said to have given birth to Apollo and Artemis under a tall palm tree. The original had been replaced by a bronze replica by Nicias in the late fifth century. The base of this dedication with Nicias's name inscribed on it can still be seen there.
p.84, para. 2 **mōns Cynthius**: the so-called mountain is 368 feet high but looms prominently over the rest of the tiny island.

★ p.84: the stone lions, visible at the back of the picture, stand beside the Sacred Way which led from the Sanctuary of Apollo to the Sacred Lake. Carved from Naxos marble at the end of the seventh century BC, they are badly weathered. Five remain *in situ*; at least one more is now in the Arsenal in Venice.

★ p.85: down this road, which leads to the Magnesian gate, went the procession at the festival of Artemis.

p.85, last para. l.1 **optiō**: the **optiō** was the adjutant to the centurion. He took over if the centurion fell. His major concern was probably with training. We have glossed the word 'sergeant', feeling that this is more appropriate than 'adjutant'.

Exercise 8.4

The centurion Lucilius is borrowed from Tacitus, *Annals* 1, 23. In the mutiny of the legions stationed in Pannonia, he was one of the first victims of the mutineers: 'centuriō Lūcīlius interficitur, cui mīlitāribus facētiīs vocābulum "cedo alteram" indiderant, quia frāctā vīte in tergō mīlitis alteram clārā vōce ac rūrsus aliam poscēbat.' (The centurion Lucilius was killed, to whom with a soldier's humour they had given the nick-name "Another please", because, when he had broken his staff on a soldier's back, he would call for another and again yet another.')

vītis (a vine-rod) was the centurion's mark of office and also his instrument of punishment. As the passage is somewhat long we have not asked for a translation. Questions 1, 4 and 5 are rather more demanding than usual, but we think the characters are drawn clearly enough to make them viable.

Background

A valuable book for further exploration of the background material in this and the following chapter is Peter Connolly's *Greece and Rome at War*, Macdonald. This is both highly informative and beautifully illustrated.

★ p.89: the soldier in our illustration has a moustache. This is an unauthentic touch! See l.6 on this page.

★ p.90: our illustration from Trajan's column shows a field dressing station. On the right, a dresser, holding a roll of bandage, attends an auxiliary with a wound in his thigh; in his pain, the soldier grits his teeth and clutches the rock he is sitting on. On the left, a medical officer examines a legionary soldier.

Chapter 9

Ablative absolute

The ablative absolute is one of the Latin constructions which are hard to put across. The cartoons should help here. Although **orīrī** and **occidere** have not been learnt, their meaning is apparent, and pupils should be encouraged from the start to translate the participial phrase by an English clause, 'as the sun was rising'. Various ways of translating e.g. **armīs indūtīs** should be offered: 'after putting on their arms', 'when they had put on their arms', 'they put on their arms and hurried'; 'arms having been put on' may be an acceptable step towards understanding the Latin but will not do as a translation.

Quīntus in Asiā

For Horace's life in the army, see *Odes* 2, 7, 6–7:

> Pompeius, the best of my friends, with whom I often broke the lingering day with wine, with garlands on my shining hair and Syrian perfume . . .

p.94, para. 1: behind Brutus's exhortation, there lies Anchises's advice to Aeneas, and to all Romans (*Aen.* 6, 851, 3):

> Rōmāne, mementō . . .
> parcere subiectīs et dēbellāre superbōs.

p.94: the sack of Xanthus is described in Plutarch's *Marcus Brutus*, which we have followed in this narrative. Plutarch dilates on the humanity of Brutus and his distress at the fate he brought on the citizens of Xanthus.
p.94, para. 2: we assume that this was Quintus's first taste of action.

★ p.95: this statue, known as the Ludovisi Gaul, is a copy of a bronze original of the late third century BC. The arms of the warrior and the left arm of his dead wife are modern reconstructions. (Museo Terme, Rome.)

Background

★ p.97: in this relief from Trajan's column, the Roman legionary soldiers who are building the camp all have their helmets off. At the back three officers supervise the work. To the left is a wood where a sentry, whose shield can be seen, is keeping guard.

p.98, para. 1 and models: the **tormentum** could lob a 60-pound missile as far as half a mile; the **catapulta** had a range of around 1,400 feet. The **ballista** could hurl 12-foot flaming darts 2,000 feet or further.

★ p.98: our illustration from Trajan's column shows the **testūdō** formation. The enemy are rushing back into their fortress, while the legionaries ascend the hill under the cover of their shields.

Chapter 10

The cartoons illustrate the use of ablative absolute with deponent and active verbs.

Quīntus tribūnus mīlitum

p.99: we do not know when Horace was promoted, but he states in *Satires* 1, 6, 48 that he was not only a tribune but also acting legionary commander: 'everyone used to get at me, a freedman's son, **quod mihi pārēret legiō Rōmāna tribūnō**.'

p. 99, last para.: Brutus and Cassius met at Sardis in August 42 BC and set out to meet the forces of Octavian and Antony.
p.100: Brutus's **daemōn malus**. The story is taken from Plutarch, as is the story of the eagles perching on the leading standards (p.100).
p.101: the first battle of Philippi took place on 23 October, the second battle about three weeks later.

Exercise 10.9

p.105, last para. **tē ipsum dēsignāvī quī legiōnī imperēs**: I have appointed you to command the legion. Relative with the subjunctive expressing purpose, not explained, but the context makes the meaning clear.
p.106, para. 1 **si moriēmur, līberī erimus**: death will bring freedom, an appropriately Stoic apophthegm.
p.106, last para. **relictā nōn bene parmulā**: quoted from *Odes* 2, 7, 10. It is doubtful whether Horace, as a tribune, even carried a shield; if he did, it would have been a **scūtum**, not a **parmula**, which means 'a little round shield' (a derogatory diminutive – 'my poor little shield'). He is echoing earlier poets who claim to have lost their shields in battle (Archilochus, Alcaeus and Anacreon) and Roman readers would have recognized this. To throw away one's shield on the field of battle was the traditional mark of a coward. Horace should not be taken literally, though we have done so in our narrative.

Background

We may have given an excessively idealized view of Brutus in our story and in this background passage. When Cicero was governor of Cilicia, he found to his disgust that Brutus was trying to wring interest at 48% out of the people of Salamis in Cyprus, to whom he had made a substantial loan. (The traditional rate was 12%.) His agent starved five of the local council to death. Caesar used to say of Brutus, 'Quidquid vult, valdē vult' (When Brutus wants something, he really wants it). When seen in the context of the Salamis affair, these words have a sinister ring to them.

Even so, he surely possessed a fundamental integrity. Quintilian's comment on Brutus's philosophical writing may well have a wider application: 'sciās eum sentīre quae dīcit' (You can tell he means what he says).

★ p.106: this coin was issued by the assassins of Julius Caesar to celebrate the killing of the dictator. The cap of freedom in the centre of the coin is a **pileus**, a felt cap worn by freed slaves. The inscription reads EID. MAR. ('the Ides of March'). (The E in EID is the Old Latin E. It had largely been dropped by Caesar's time.)

Chapter 11

There is no new grammar or syntax in this chapter and we have again devoted the cartoons to illustrating the use of participles.

Quīntus in Italiam redit

p.109, para. 2 **frācta est virtūs . . . tetigērunt**: a paraphrase of *Odes* 2, 7, 10-11:

> With you I experienced Philippi and swift flight, leaving my poor shield behind dishonourably, when Virtue was broken and those who had threatened (Antony) touched the ground in shame with their chin.

Brutus, the Stoic, and his cause represent Virtue.
p.110, para. 3: the date of Horace's return to Italy is not known. There was an amnesty in 39 (Peace of Misenum), which allowed exiles to return, but he may well have come back earlier and remained inconspicuous. This and the following chapter are intended to build up a picture of the terrible misery and disruption caused by the Civil War. Horace describes his return in *Epistles* 2, 46ff.

> But harsh times moved me from the place I loved (i.e. the Academy), and the tide of civil war carried me, a novice, against the arms of Augustus Caesar, for which my muscles were no match. As soon as Philippi sent me off from there (i.e. from life in the army), humbled, with wings clipped, robbed of my father's home and farm, poverty, which makes a man rash, drove me to write verses.

p. 110, para. 3 **quod . . . vidēret**: subjunctive, because the **quod** clause is part of Quintus's thoughts (sub-oblique).

p.111, para. 2 **revēnī ut . . . revīsam**: 'I have come back to visit' (primary sequence).
p.111, para. 3: on his return to Italy, Octavian had to demobilize and settle 100,000 veterans. To provide for them he confiscated land from eighteen Italian towns which had not supported his cause enthusiastically; among these towns were Venusia and Mantua. Horace's father, if he was still alive, was dispossessed, as Virgil's father may have been at Mantua.
p.111, para. 3, l.5 **adiūvissent**: subjunctive, because the **quae** clause is part of Octavian's point of view, i.e. sub-oblique.

Exercise 11.4

p.113, para. 1: compare Virgil, *Georgics* 1, 502ff.:

> Too long already has the court of heaven begrudged us your presence, Caesar, and complained that you care for triumphs among men; for here on earth right and wrong have been turned upside down; there are so many wars throughout the world, so many forms of wickedness; the plough has none of the honour it deserves; the fields lie neglected when the farmers have been taken away from them; and curved pruning hooks are beaten into straight, unbending swords.

Background

Octavian was born on 23 September 63 BC. He was in Illyricum when Caesar was murdered in March 44, and, after a brief hesitation, he returned to Italy. What were the feelings of this eighteen-year-old when he discovered on landing near Brundisium that he was Caesar's heir? Did he resolve then and there to become the most powerful man in the Western world? ('Look at his name; then look at his age.' – Cicero, *ad Att.* 6, 8)
p.115: Virgil's estate. We have followed the traditional view that Virgil, after losing his estate in the confiscations, won it back through a successful petition. Some modern scholars are less inclined to believe that *Eclogue* 1, from which this is inferred, is autobiographical: Virgil could in fact be conveying in a general way the consequences of the land confiscations.
p.115, last para.: a new Golden Age. See Virgil, *Eclogue* 4, written in 40 BC. This famous poem tells how:

> ultima Cūmaeī vēnit iam carminis aetās;
> magnus ab integrō saeclōrum nāscitur ōrdō.

(The final era foretold by the Sibyl of Cumae in her song has now come; the great cycle of ages is born anew.)

★ p.115: Mother Earth sits holding children in her arms amid fruit, flowers and animals. This relief from the Ara Pacis (13–9 BC) captures the spirit of Virgil's dream of another Golden Age.

Chapter 12

cum = when (in past time)

Other temporal conjunctions (**ubi, postquam, simul ac** etc.) are followed by the indicative, as in English, and the grammarians produce no convincing reason why the subjunctive is used with **cum** in clauses which appear to be purely temporal. This may not bother your pupils, but you will have to stress that these subjunctives cannot be translated by English subjunctives. The usages of **cum** are in fact most complex and in many cases it is followed by the indicative even in past time (see Part III, Chapter 18).

Indirect questions

p.117: these occasion little difficulty once pupils understand that Latin uses the subjunctive, whereas English uses the indicative. It will gradually become apparent that Latin uses the subjunctive in all clauses (except indirect statement) which are oblique or sub-oblique. This point will be made in Part III, where relative and **quod** clauses with the subjunctive are explained.
The other difficulty in indirect questions is that Latin uses this construction in all subordinate clauses introduced by an interrogative word, e.g. I do not know when he will come: **nesciō quandō ventūrus sit**; or I told him what I had done: **eī dīxī quid fēcissem**. To our minds there does not appear to be a question in such cases. But this will probably not cause much trouble in translating from Latin into English.

Quīntus Rōmam redit

p.119, penultimate para.: Quintus's dream. Horace in several odes claims that he is under the special protection of the gods, e.g. *Odes* 1, 17:

> dī mē tuentur, dīs pietās mea
> et mūsa cordī est.

(The gods watch over me, my piety and my muse are pleasing to their hearts.)

★ p.120: this picture on a wine-jar of the fifth century BC shows Apollo playing the cithara and holding out a dish in his right hand. On the right, Hera holds a jug. On the left stands the winged figure of Nike. (British Museum).

p.120, para. 2: Sextus Pompeius, son of Pompey the Great, carried on the struggle against Octavian from Sicily; he had a large fleet and blockaded the Italian coast. He supported Antony against Octavian in 40; made a treaty with the Triumviri in 39 (Treaty of Misenum); was soon at war with Octavian again; and defeated him in two sea battles in 38. In 37 Antony gave Octavian 120 ships (Treaty of Tarentum) and the following year Agrippa defeated Pompeius at Naulochus. He was captured and executed in 35.
p.120, para. 4 **clēmentiam**: by the terms of the Treaty of Misenum (39), exiles who had supported Pompeius were allowed home. Octavian had respected Marcus's father, although he had consented to his death. He made Marcus his colleague as **cōnsul suffectus** in 30 BC and proconsul of Asia the following year.

Marcus's appointment as **quaestor aerārius** is fictional. Horace did in fact secure the post of **scrība quaestōrius**, when he returned to Rome, cf. Suetonius, *Vīta*: 'Involved in the Philippi campaign by Marcus Brutus the general, he served as a military tribune; and when his side was defeated, he obtained a pardon and secured a secretarial post in the Treasury.' **comparāvit** (secured) suggests that he may have bought the post.
p.121, last para.: the lines quoted are from *Epode* 2, 1–28, with omissions. There are seventeen epodes, written between 41 and 31 and published 30 BC. Epode 2 is mildly satirical; the list of the joys of country life is conventional and similar lists occur in other Augustan poets; Virgil, *Georgics* 2, 458ff., quoted in the next chapter, has elements in common with it. But Horace attributes these thoughts to Alfius the money lender, who is always about to retire to the country but never does so.

Characteristically, Horace disguises the direction of his thought until the end, when the last four lines put quite a different complexion on what came before. There is a strong note of irony in l. 6 (**solūtus omnī faenore**), attributed to Alfius.

The lines quoted are linguistically within the compass of pupils who have reached this point.

The subject matter has a certain appeal and raises echoes of sympathy today and the twist at the end raises a smile; you might ask your pupils to put similar thoughts into the mouth of a contemporary merchant banker or stockbroker.

As we say in the pupil's note, the main difficulty will be in the word order, especially the separation of adjectives from the nouns they agree with; the only solution to this problem is for them to watch word endings with particular care and analyse correctly.

The metre of this poem is extremely simple (iambic trimeter followed by iambic dimeter); if you are going to teach your pupils scansion, this would be a far easier starting point than dactylic hexameters or elegiac couplets. See Appendix on metres.

★ p.122: this panel, showing an idyllic landscape, is from a wall-painting in the villa of Agrippa Postumus at Boscotrecase (about 10 BC).

Exercise 12.6

This passage is suggested by the lines in *Satires* 1, 6, 45–8:

> Now I return to myself, the freedman's son, whom everyone disparages as the freedman's son, now because I am your friend, Maecenas; but once because a Roman legion obeyed me as a tribune.

Although Horace became the close friend of Maecenas and other leading men, and eventually of Augustus himself, he undoubtedly encountered much snobbery earlier in his career.
p.124, last para. **quī cum senātōre rem ageret**: 'to do business with the senator'; relative with the subjunctive expressing purpose.

★ p.124: the photograph shows the remains of the **tabulārium** at Rome, the building where the state archives were kept. It was designed in 78 BC. The podium and gallery are original. The upper storey was removed to make way for the medieval senatorial palace.

Background

This thumb-nail sketch of the development of Latin poetry is primarily intended to put Horace into a context which will help to make his poetry more easily intelligible. It is grossly oversimplified but it has still been impossible to keep out a lot of unfamiliar material, especially in the references to Greek poetry. Pupils should not be expected to learn all the facts given but it may help to prepare them for some of the unexpected features of ancient poetry.

In our selection we have tried to choose poems and passages which have an immediate appeal even though the genesis and background of the poems is not understood. We have introduced Horace's poetry in the context of our narrative, a context which is sometimes historically sound, e.g. the journey to Brundisium (*Satires* 1, 5), at other times fictitious but appropriate, e.g. in Chapter 16 we make him write the **Fōns Bandusiae,** *Odes* 3, 13, the day he has first seen his Sabine farm. As we shall see when we come to it, the genesis of the latter ode is extremely complex, but little would be gained by trying to put our pupils at this stage into the full picture, even if that were accessible.

★ p.126: the fresco is from Pompeii and is now in the Museo Nazionale, Naples. The man carries a papyrus scroll. His wife holds a stylus to her lips and has in her left hand a two-leaved wooden tablet spread with wax.

Chapter 13

Indirect questions 2

Indirect questions involve the use of the perfect subjunctive and the periphrastic future. This completes the tables of the subjunctive mood. We consider that the table on p.129 sufficiently illustrates their use without the need for a full statement of the rules of sequence in the traditional form. Sequence causes no difficulty in reading Latin except in the rare instances where the true perfect is followed by primary sequence, e.g. **rogāvērunt quid factūrus sīs** They have asked what you are going to do.

Pupils may have difficulty in English-Latin sentences involving future subjunctives, e.g. Exercise 13.5.1; this can be solved by using a periphrastic form of the English: The boy asked when his father would set out = The boy asked when his father *was* (**esset**) going to set out. I do not know when we will set out = I do not know when we *are* (**simus**) going to set out.

Quīntus amīcīs veteribus occurrit

p.131, para. 1 **campum**: the Campus Martius, see Part I, Chapter 17.
p.131, para. 3 **Varius**: Varius Rūfus, a close friend of Virgil and himself a poet. Horace praises him as the leading epic poet. (*Satires* 1, 10, 43–4 **forte epos ācer/ut nēmō Varius dūcit**). After Virgil's death, on Augustus's orders, he prepared the *Aeneid* for publication.
p.131, para. 4 **quid dē novō statū rērum sentīrent**: 'the new state of affairs' refers to Octavian's **clēmentia** and the dawning hopes of peace and stability.
p.131, para. 5 **dē rē rūsticā scrībēbat**: Virgil wrote the *Georgics* between 37 and 30 BC; we have made him start them a little earlier.
p.131: the extract is from *Georgic* 2, 458 ff., a great paean of praise for the virtues of country life, contrasted with the life of the city, which is characterized by artificiality, greed, luxury and lust for power. It is harder than the extract from Horace and so we have first given a prose paraphrase in which the words have been rearranged in a more familiar order. The passage may still prove too hard for all but the ablest pupils; if so, don't get bogged down in it. We give C. Day Lewis's translation, including (in brackets) the lines we have omitted and the conclusion of the paragraph:

> Oh, too lucky for words, if only he knew his luck,
> Is the countryman who far from the clash of armaments
> Lives, and rewarding earth is lavish of all he needs!
> True, no mansion tall with a swanky gate throws up
> In the morning a mob of callers to crowd him out (and gape at
> Doorposts inlaid with beautiful tortoiseshell, attire

Of gold brocade, connoisseur's bronzes.
No foreign dyes may stain his white fleeces, nor exotic
Spice like cinnamon spoil his olive oil for use):
But calm security and a life that will not cheat you,
Rich in its own rewards, are here: the broad ease of the farmlands,
(Caves, living lakes, and combes that are cool even at midsummer,)
Mooing of herds, and slumber mild in the trees' shade.
(Here are glades game-haunted,
Lads hardened to labour, inured to simple ways,
Reverence for God, respect for the family. When Justice
Left earth, her latest footprints were stamped on folk like these.)

l.1 **nōrint** = **nōverint**: perfect subjunctive.
l.4–5 **māne salūtantum**: a vast wave of early-morning callers. This refers to the custom of clients calling at the houses of their patrons to escort them to the forum; the larger the crowd of clients escorting a man, the more important he would appear.
l.7 **mūgītusque boum** and the lowing of oxen (or cows, if you prefer).

★ p.132: this work by the great French seventeenth-century landscape painter Claude Lorraine captures the spirit of the passage from the *Georgics* which we have quoted. (Barber Institute of Fine Arts, Birmingham)
p.132, para. 3 **ad Palātium**: Palātium means the Palatine hill. Augustus lived there, first taking over the house where Hortensius had lived, and then building a temple to Apollo and a magnificent new house (28 BC); it was this house which came to have the meaning 'palace'.

p.134: *Odes* 2, 7. It is early in the course to introduce an ode of Horace but the narrative demands that it comes here; some of the difficulties have been anticipated in the narrative and it occurs in a known context which is a great help to understanding. The main obstacle will again be word order and so we give a prose paraphrase before the actual poem. The following points may require explanation:
ll.1–2 **O Pompei** and **dēducte** agree (vocative case).
l.2 **Brūtō . . .**: ablative absolute = when Brutus was leader of our army.
l.3 **Quirītem: Quirītēs** was the old name for the Roman citizens in a civilian context; the word therefore implies both that Pompeius has recovered full citizen rights and has left the army. **quis tē redōnāvit?** the question might be answered 'Jupiter' (see l.13), or 'Octavian' (who had pardoned him).
prīme first = best, closest.
l.5 **frēgī** I broke (the back of) the day: the use of this word suggests that the day was boring and resisted attempts to hurry it on.
l.6 **corōnātus**: perfect participle used like a Greek middle – 'having garlanded my hair'; in fact he did not garland his hair with perfume but put both garlands and perfume on his hair. The custom of wearing garlands and applying lavish amounts of perfume at dinner parties may need comment.
ll.6–7 **Philippōs et celerem fugam**: best taken as a hendiadys – 'swift flight at Philippi' (but this point may be glossed over with your pupils at this stage).
l.7 **parmulā nōn bene relictā**: see page 106 and note.
l.7 **virtūs frācta**: courage, i.e. brave men, were broken, but the cause of the Stoic Brutus was also the cause of Virtue, so the phrase is ambivalent.
l.8 **solum mentō . . . tetigēre** touched the ground with their chin: this could either mean 'bit the dust (in death)' or that after the battle they bowed low to their conquerors (this gives a better contrast with **'minācēs'**). **tetigēre** = **tetigērunt** (this form is very common in verse). **turpe**: this is best taken, in our view, as an adverb, 'dishonourably'; it could agree with **solum** – the base earth.
ll.9–10 **Mercurius mē . . . sustulit**: Horace implicitly compares himself with an Homeric hero rescued from battle by the gods, e.g. in *Iliad* 3,380ff. Aphrodite rescues Paris when he is in danger during his duel with Menelaus: 'Aphrodite snatched him away, easily, as a goddess can, and covered him with a thick mist and set him in his sweetly smelling chamber.' Roman readers would immediately catch the reference and appreciate the ironical tone (reinforced, we think, by **paventem**; his aerial flight was an alarming experience).
l.10 **tē unda . . . resorbēns**: the image is that of a man struggling to reach the safety of the shore, sucked out again to sea by the currents.

aestuōsīs seething, boiling, a word evocative of dangerous currents.
ll.11–12 **Iovī dapem obligātam** the feast owed to Jupiter. This implies that the party will begin with a sacrifice to Jupiter, vowed as a thank-offering for Pompeius's safe return.
l.13 **sub laurū meā**: the party is taking place in Horace's garden. In the lines omitted the pace quickens; Horace gives a series of staccato orders for starting up the party. The poem has developed into a dramatic monologue (see commentary on **vīxī puellīs**, Chapter 19).
ll.13–14 **cadīs tibi dēstinātis**: Horace has marked some of the jars in his cellar: 'For Pompeius's homecoming.'

Questions 3 and 4: Horace belittles his own performance by saying that 'he left behind his poor shield dishonourably', which we do not think can be taken literally. His escape is described in mythical terms with a note of irony and humour. On the other hand, Pompeius's fate is described seriously in a striking image. He therefore at once laughs at his own performance and sympathises with Pompeius's prolonged struggle in the storms of war.
Question 8: his feelings towards Pompeius include strong affection (**prīme sodālium**), expressed with considerable emotional force (the language of the opening stanzas is elevated and rhetorical); he recollects dangers faced together and good times enjoyed. He has reserved some wine for the occasion and is happy to run riot **receptō amīcō**.

We append a translation of the whole ode by James Michie (1964):

Pompeius, chief of all my friends, with whom
I often ventured to the edge of doom
When Brutus led our line,
With whom, aided by wine

And garlands and Arabian spikenard,
I killed those afternoons that died so hard –
Who has new-made you, then,
A Roman citizen

And given you back your native gods and weather?
We two once beat a swift retreat together
Upon Philippi's field,
When I dumped my poor shield,

And courage cracked, and the strong men who frowned
Fiercest were felled, chins to the miry ground.
But I, half dead with fear,
Was wafted, airborne, clear

Of the enemy lines, wrapped in a misty blur
By Mercury, not sucked back, as you were,
From safety and the shore
By the wild tide of war.

Pay Jove his feast, then. In my laurel's shade
Stretch out the bones that long campaigns have made
Weary. Your wine's been waiting
For years: no hesitating!

Fill up the polished goblets to the top
With memory-drowning Massic! Slave, unstop
The deep-mouthed shells that store
Sweet-smelling oil and pour!

Who'll run to fit us out with wreaths and find
Myrtle and parsley, damp and easily twined?
Who'll win the right to be
Lord of the revelry

By dicing highest? I propose to go
As mad as a Thracian. It's sheer joy to throw
Sanity overboard
When a dear friend's restored.

Exercise 13.7

This passage is partly based on the end of *Satires* 1, 6, where Horace describes his modest and quiet life in Rome, ending (ll.128 ff.):

This is the life of men free from the burden of wretched ambition; I comfort myself that I shall live more pleasantly like this than if my grandfather and father and uncle had been quaestors.

Background

p.138, para. 3: Catullus calls his newly printed collection of poems 'lepidum novum libellum/ āridā modo pūmice expolītum' (1,1–2) (a delightful, novel little book, freshly polished with dry pumice). Pumice was used to polish the ends of the rolls of papyrus.

★ p.138: back row: four leaves of a wooden writing tablet; inkpots in faience, pottery and bronze.
front row: a letter in Greek on papyrus; (from the front) a reed pen, a bronze pen, an ivory stylus and a bronze stylus.

p.139, para 2. **volūmen**: cf. English word 'volume'.

Further reading: U.E. Paoli: *Rome, Its People, Life and Customs*, Longman, pp.174–90; *Cambridge History of Classical Literature*, Vol. 1 (ed. P. E. Easterling and B. M.W. Knox), pp.7–22, & Vol. 2 (ed. E. J. Kenney and W. V. Clausen), pp.15–27.
★ p.139: the manuscript, which dates from 20 BC, is the oldest Roman book in existence. The work of the poet Gallus (see Teacher's Book p.48), the four most legible lines read:

FATA·MIHI·CAESAR·TVM·ERVNT·MEA·DVLCIA·QVOM·TV
MAXIMA·ROMANAE·PARS·ERIT·HISTORIAE·
POSTQVE·TVVM·REDITVM·MVLTORVM·TEMPLA·DEORVM
FIXA·LEGAM·SPOLIEIS·DEIVITIORA·TVEIS

My fate, Caesar, will only then be sweet to me, when you are the most important part of Roman history, and when after your return I read how the temples of many gods have been made richer by your spoils fixed up in them.

The Caesar referred to is, of course, Octavian.

The archaic spellings: **quom = cum; spoliēīs = spoliīs; dēīvitiōra = dīvitiōra; tuēīs = tuīs.**

Chapter 14

Indirect statement

The cartoon captions should prove particularly useful in introducing this construction, which, at first appearance, is difficult. A great deal of information has to be acquired simultaneously – the full range of the infinitives, the use of the reflexive pronouns etc. – and it is not easy to simplify.

Exercise 14.1

4 and 5: difficulty may be caused by the succession of nouns in the accusative. The subject of the infinitive usually comes before the object; we always use this order in our examples.

Exercise 14.4

The tenses of the infinitives to be supplied, e.g. in 5 **bibere** (were drinking) and **bibisse** (had drunk) would both be correct answers.

Quīntus Maecēnātī commendātur

Horace's introduction to Maecenas, usually dated to 38 BC, is described in *Satires* 1, 6, quoted later in this chapter.
p.146, para. 2 **dēfuisset . . . praebuisset**: subjunctives because the verbs are virtually oblique, i.e. they express what Horace felt, not facts vouched for by the author of the narrative. This should be glossed over unless questions are asked.
p.146, para. 2 **occupātum esse . . . abesse** etc: infinitives because they are part of what Virgil said; in continuous indirect statement subject pronouns are often omitted.
p.146, bottom: *Satires* 6, 45ff. with omissions. In this satire Horace praises Maecenas for his lack of snobbery, censures those who consider birth more important than character, and defends his own position as a freedman's son who has become a friend of Maecenas.
l.3 **quia sim: quia** normally takes the indicative for factual reasons but here the reason is expressed from the point of view of those who run Horace down (so also **pārēret** in next line). Some of the difficulties of this passage, especially in vocabulary, have been anticipated in the preceding narrative.
l.12 **quod eram nārrō** I say what I (really) am: the neuter referring to persons is a common idiom, cf. p.142 **dīximus quid essēs.**
l.14 **magnum hoc ego dūcō: hoc** is accusative neuter singular (I consider this a great thing . . .). The vowel sounds short but the syllable in the Augustan poets is regularly heavy; W. S. Allen: *Vox Latina*, p.76, explains this on the grounds that the neuter singular was originally **hocc** (the ablative on the other hand is **hōc** with a long vowel, cf. l.5 above).
l.15 **quī turpī sēcernis honestō** who distinguish the honourable (man) from the base (man). The gender of the adjectives could be either masculine or neuter.
l.16 **nōn patre praeclārō** (I pleased you) not by

(= because of) a famous father . . . .
ll.19–20 **magnī quō puerī magnīs ē centuriōnibus ortī**: the big boys would have been the sons of veterans settled in the colony by Sulla shortly before Horace was born; these were now the big men of Venusia. **magnī** is ambivalent in meaning, as is English 'big'.
p.147, l.5 **artēs quās doceat** the (sort of) arts which any knight and senator has his own children taught. **doceat** is probably generic subjunctive. **doceat** appears to be used causatively, like the Greek *didaskomai*.

It is worth continuing the quotation from *Satire* 6 in which Horace pays a famous and moving tribute to his father:

> If anyone had seen my clothes and the slaves following me, in the throng, he would have believed their cost was supplied by an ancestral estate. My father himself, an incorruptible guardian, was there at all my teachers'. In short, he kept me pure, which is the crown of virtue, not only from every wrong deed but also from foul slander. Nor was he afraid that someone might at some time hold it against him if I followed a low-paid profession, as an auctioneer, or an auctioneer's agent, as he was himself; nor would I have complained; and now, because of this, I owe him all the greater praise and gratitude. I would never be ashamed of such a father, as long as I were in my right mind.

The narrative of this chapter gives ample scope for discussion of several topics – snobbery, the relations between Horace and Maecenas and between Horace and his father; did his father do the right thing in taking Horace to Rome, leaving his wife (as we presume) in Venusia (Horace never mentions his mother and she may, of course, have been dead)?

Exercise 14.7

A straightforward passage, preparing the way for *Satires* 1, 5. Anxur, better known by its later name, Terracina, where Horace was to meet Maecenas and Cocceius, was at the end of the canal which ran from Forum Appi; see map (p.152). Sinuessa, where they met Virgil and Varius, was forty miles further along the Appian way.
Maecenas's mission: Octavian had sent Maecenas on an embassy to Antony at Brundisium in 40 BC, which led to the so-called Peace of Brundisium. This cannot be the occasion of our journey, since Maecenas had not met Horace then. In 38 Octavian again sent Maecenas to negotiate with Antony, in Athens. This time his mission led to a meeting between Octavian and Antony the following spring at Tarentum, where they patched up their differences and Antony lent Octavian 120 ships, which enabled him to defeat Sextus Pompeius.

Background

★ p.149: this fresco of a fruit and flower garden is from the Imperial Villa of Livia, the wife of Augustus, at Prima Porta near Rome. It was restored in 1952–3 just before falling into complete decay and is now in the Museo Nazionale at Rome. Bernard Berenson wrote: 'How dewy, how penetratingly fresh are grass and trees and flowers, how coruscating the fruit. Pomegranates as Renoir painted them. Bird songs charm one's ears. The distance remains magically impenetrable.'

p.149, para. 2 **Scrībōnia**: Octavian married her in order to improve his relations with Sextus Pompeius whose sister-in-law she was, but the match was not a success. Octavian divorced her on the very day that she bore him a daughter.
p.150, para. 4: from the tall tower of Maecenas's house on the Esquiline Hill, the Emperor Nero looked down on Rome as it burnt.

As time went by, a coolness developed between Augustus and Maecenas. This may have been due to a love affair between Augustus and Maecenas's beautiful but difficult wife, Terentia. However, when he died childless in 8 BC, Maecenas bequeathed his considerable property to his former friend.

Chapter 15

Quīntus Brundisium iter facit

pp.151–3: *Satires* 1, 5, 1–29.
p.151, l.1: the quantity of the **-ās**, marked by macra, shows that **magnā** agrees with **Rōmā** (ablative) and **Arīcia** must be subject.
p.152, l.2: bad water was a perpetual hazard for travellers in the ancient world and, indeed, in Italy during this century.
l.4 **iam nox** . . .: the language becomes more

elevated, assuming a mock heroic style in striking contrast with the colloquial language which follows; your pupils might notice this with some encouragement.
p.152, l.7 **ingerere**: historic infinitive, 'began to hurl'.
l.7 **trecentōs**: 300 is used for an indefinitely large number.
l.9 **malī culicēs**: the canal ran along marshy ground, notorious for its mosquitoes.
l.13 **missae pāstum . . . religat**: difficult word order; simplified, it would read **nauta piger retinācula mūlae pāstum missae saxō religat**. Analysis will give the answer. **pāstum** is supine in **-um**, expressing purpose.
p.153, l.4: Feronia was an ancient Italian goddess, consort of Jupiter of Anxur. The sudden apostrophe of this obscure goddess will require explanation. It is not made any easier by the fact that her spring appears to lie three miles short of the town of Anxur. Apostrophe is a figure of high poetry, here used ironically. For a serious example of apostrophe, see Virgil *Georgics* 2, 146–8, where Virgil, in a rhapsody on the beauties of the Italian countryside, apostrophizes the famous springs of Clitumnus:

> hinc albī, Clītumne, gregēs et maxima taurus
> victima, saepe tuō perfūsī flūmine sacrō,
> Rōmānōs ad templa deum dūxēre triumphōs.

(from here the white flocks and the bull, the greatest sacrifice, often sprinkled with your sacred water, Clitumnus, have led the Roman triumphs to the temples of the gods.)
l.6 **saxīs lātē candentibus**: Anxur (Terracina) is on top of a hill of white limestone.
l.9 **āversōs solitī compōnere amīcōs** accustomed to bring together friends who had fallen out. This refers to their mission of 40 BC (see page 149).
l.13: Plotius, another close friend of Virgil, helped Varius edit the *Aeneid* after Virgil's death. The three friends meet at Sinuessa, which is on the coast.

Indirect command

This construction causes little trouble, except the use of **iubeō** and **vetō** + infinitive, when translating from English into Latin; the Romans did not write, e.g. **iubeō tē nōn hoc facere** but **vetō tē hoc facere.** We do not give examples of this from English into Latin.

Exercises 15.5

We revert to Cicero in this exercise, using a letter which illustrates travel by sea, since the background section is on travel. Pupils should be warned of the time slip. Cicero was returning from Cilicia, where he had been governor, in November 50 BC. He writes from Brundisium, having left Tiro at Patras. Cicero's dependence on Tiro for help in every sphere is clear from the fact that he wrote eleven letters to him while they were parted between 3 November and the end of January. All show great anxiety for Tiro's health and keen concern that Tiro should rejoin him as soon as possible. This letter is *ad Fam.* 16, 9.

★ p.157: from a marble relief of about 200 AD, found at Ostia. The emblem of Rome (the wolf and twins) is twice depicted in the sails. Flames curl up from a lighthouse at the rear. To the right stands the god Neptune with his trident.

Question 1: Cicero arrived at Brundisium on 25 November at the fourth hour, i.e. about 10.30 a.m. He had been sailing for two nights and a day, leaving on 23 November after dinner.
Question 3: the letter was dated 13 November and arrived on the 27th.

Background

Further reading: Paoli: *Rome, Its People, Life and Customs*, pp.228–31; O. A. W. Dilke: *The Ancient Romans*, David and Charles, Chapter 4. The Roman roads in Britain are particularly straight. If there is one near you, you can discuss it with your pupils.

★ p.159: this relief of a **raeda** comes from Klagenfurt.

p.160: If your pupils draw a cross-section of a Roman road, it would probably be helpful to encourage them to include a ditch on either side of the road rather than to reproduce the kerb in the illustration on p.158.

Chapter 16

This chapter introduces three linguistic points:
1 use of **negō**

2 **fīō** + complement
3 **opus est** + ablative;
the first two are illustrated in the cartoon captions.

Quīntus rūsticus fit

ll.2–3 **carmina multa scrīpserat**: in this period Horace was engaged on *Satires*, Book 1, and the *Epodes*.
ll.5–6 **paucī . . . dīcēbant sē poētās antīquōs mālle**: compare *Epistles* 2,1, 50ff. (written much later, after his *Odes* had had a disappointing reception):

> Ennius is wise and brave and a second Homer, the critics say . . . isn't Naevius always in their hands and sticks deep in their minds as if he were quite recent? . . . I am indignant that anything should be faulted not because it is considered clumsily or inelegantly written but because it is recent; and that they don't ask for indulgence for the ancient writers but prizes of honour.

p.161 para. 3: *Satires* 1, 9, 1–21.
poem, l.2 **nescioquid nūgārum: nescioquis** is used idiomatically as a pronoun, meaning 'someone or other'. Here the neuter is followed by a partitive genitive, compare **aliquid cibī** = some food.
l.2 **tōtus** = wholly involved in.
p.162, l.1 **inquam**: I say (the only other forms of this verb are **inquit** and more rarely **inquimus, inquitis, inquiunt**).
l.3 **nōrīs nōs: nōrīs** = **nōverīs; volō ut nōs** (= **mē) nōverīs**: I want you to make my acquaintance.
l.3 **doctī sumus: doctus** (learned), is often used of poets: **sumus** is probably plural for singular, like **nōs**: 'I am a poet'.
ll.3–4 **plūris hōc mihi eris** you will be worth more in my eyes because of this. **plūris** is genitive of value, **hōc** ablative of cause.
ll.5–6 **īre . . . cōnsistere . . . dīcere**: historic infinitives. In this passage Horace's innate courtesy forces him to reply with cold politeness to the bore's outrageous behaviour. He cannot bring himself to be rude, as Bolanus would have been (an unknown character, evidently notorious for his quick temper).
l.11 **iamdūdum videō** I've seen this for a long time (present of remaining effects, cf. French 'je suis ici depuis longtemps').
l.14 **Caesaris hortōs**: these gardens on the Janiculum hills near the Tiber had been left to the people by Julius Caesar as a public park.
l.15 **nīl habeō quod agam** I have nothing to do, I'm not busy. In this satire, Horace not only satirizes the bore, a vulgar, pushing type of man, but also throughout laughs at himself with characteristic irony. It is worth quoting the following passage in which the bore asks Horace to introduce him to Maecenas; it skilfully develops the insensitive character of the bore and throws a flood of light on the relations between Horace and Maecenas: ll.35–60:

> We had reached the temple of Vesta and by now a quarter of the day was past, and it happened that by chance he had to appear in court as he was held to bail; if he failed to do so, he would lose his case. 'Please,' he said, 'give me your support in court for a little.' 'Hang me if I have the strength to stand so long and I know nothing of the civil law; and I am hurrying, you know where.' 'I don't know what to do,' he said, 'whether to leave you or my business.' 'Me, please.' 'I won't do it,' he said and began to lead the way. I follow, for it is hard to fight with one's conqueror. Then he starts up again: 'How do you stand with Maecenas? He's a man of few friends and good sound sense. No one has used his luck more cleverly. You would have a great helper, who would play second fiddle to you, if you were willing to introduce me to him. Hang me, if you didn't push all the rest out of your way.' 'We don't live the way you think at Maecenas's. No house is purer than his or more remote from vices of that kind. It's no disadvantage to me, if this man is richer or more learned than me; each one of us has his own place.' 'What you say is extraordinary, scarcely credible.' 'And yet it's true.' 'You fire me to want all the more to get near him.' 'Only wish it, and such is your courage, you will take him by storm; he is a man who can be conquered and for that reason he makes the first approaches difficult.' 'I won't let myself down. I'll bribe his servants with gifts; if I'm shut out today, I'll not give up; I will find out his time-table; I'll meet him at the cross roads; I'll escort him home. Life gives nothing to mortals without great efforts.'

★ p.162: on the right the **Pōns Fabricius** (62 BC) leads to the Tiber Island, on which a temple to Aesculapius was dedicated in 293 BC. The bridge to the left, the **Pōns Cestius**, dates originally from 46 BC. It was rebuilt in 1892.

p.163, para. 2 **parvum fundum**: Maecenas gave Horace the Sabine farm, which transformed his life, in about 35 BC; for a descripton, see background section, p.170. The farm was in the valley of the Digentia (modern Licenza), a tributary of the Anio; it was near the town of Varia (modern Vicovaro). Two Roman houses have been discovered in this area, of which the one near the village of Licenza seems to fit Horace's own description (*Epistles* 1, 16) better. It was excavated in 1911 and the description we give is of this house (Exercise 16.5).

★ p.163: the remains of a villa, thought to be that of Horace.

p.164, last para.: *Satires*, 2, 6, 1–5.
ubi . . . foret where there would be: **foret** is a subjunctive formed from the future infinitive **fore.**
bene est it is well, I am satisfied.
fāxīs: archaic perfect subjunctive = **fēcerīs**

Exercise 16.5

We suggest that the background section, especially the paragraph on Horace's farm, is read before this exercise is done. It is rather long but, apart from the poem, straightforward, when the context is clearly known.
p.164, para 2, l.5 **vīnum paulum acerbum vidētur**: compare *Odes* 1, 20 (quoted in Chapter 17, Exercise 4). Falernian is one of the grandest wines.

p.167 **o fōns Bandusiae**: *Odes* 3, 13.
This is a complex ode. It takes the form of a hymn to the Fons Bandusiae. All springs were sacred and in mid-October the Fontinalia were celebrated, when offerings of flowers and wine where thrown into the water (Varro, *dē Linguā Latīnā* 6, 22). **crās** (1.3) suggests that this, or a similar ceremony, will take place tomorrow (the third verse evokes the heat of midsummer, not October, and so the Fontinalia itself is presumably not the occasion).

The formal characteristics of a hymn are: (1) The invocation – **ō fōns Bandusiae** (2) the attributes of the deity – **splendidior vitrō** (3) the functions of the deity (how he/she helps mortal creatures), introduced commonly by **tū/tē** . . . in anaphora.

But this poem has two digressions from the hymn form: (1) five lines (nearly one third of the whole poem) devoted to the kid and its fate; (2) the last stanza, introduced by **tū** in anaphora, does not describe a present function or attribute of the spring but foretells the fame it will win through Horace's poetry.

The first digression describes the kid sympathetically in ll.4, 5, and 8. **frūstrā**, both by its meaning and by its weight and by its isolation, in emphatic position at the beginning of a line between stops, sounds like a knell of doom; it is a remarkable change of tone. This is followed by a striking image in which the kid's red blood stains the cool, clear water. Is it a just response to feel a strong revulsion at the fate the kid is to suffer?

The last stanza takes a quite unexpected direction in its first line. Horace's poem will make the **fōns Bandusiae** one of the springs famous in literature, like Castalia or Arethusa. The ode ends with a naturalistic description of the scene, in which the rhythm (enjambements make the last lines seem to tumble over each other) and the sounds (see Gordon-Williams: *Horace Odes*, Book III, Oxford) assist the sense.

So, if we ask what the poem is about, the answer will be complex. Clearly, it is about the spring and its beauty; it has a dimension of 'nature' poetry. It is also about the kid and the sadness of animal sacrifice (a very unexpected note in ancient poetry). Lastly, it is about Horace's own poetry, as he claims that he can confer on the spring the gift of immortal fame. It is characteristic of Horace in his *Odes* to revitalize old forms, to pack much into a small compass, to change direction in thought, to vary the tone; all this is achieved here in a mere sixteen lines. We will see him performing a similar feat in **vīxī puellīs** (*Odes* 3, 26) in Chapter 19.
p.168, l.1 **flagrantis . . . Canīculae**: the Dog Star (the constellation called the 'Lesser Dog') rises in late July, the hottest part of summer; compare English 'dog days'.
l.2 **nescit tangere** knows not how to touch you = cannot touch you.

Background

We owe much of our evidence about Roman houses to the eruption of Vesuvius on 24

August AD 79. The town closest to the mountain, Herculaneum, was swamped in a flow of volcanic mud; Pompeii was covered with a fall of ash and stones. The excavations of these towns have proved enormously valuable. See *Pompeii AD 79*, Royal Academy of Arts; Michael Grant: *Cities of Vesuvius*, Spring Books; Paoli: *Rome, Its People, Life and Customs*, pp.54–77; *Oxford History of the Classical World*, pp.718–37. See also Pliny, *Ep.* 2, 17.

★ p.169: this peristyle garden is in the House of the Vettii at Pompeii.

★ p.170: this reconstruction is based on the detailed description of his Laurentine villa in one of the letters (*Ep.* 2, 17) of the statesman and writer, Pliny the Younger (AD *c.* 61–*c.* 112).

p.171: the 'missing room' in our plans 1, 2 and 3 is a kitchen. Kitchens had no fixed position in the normal layout of a Roman house. They tended to be poky little holes, fitted in wherever there was some free space in a building. Cooking in these cramped conditions was both difficult and dangerous.

Chapter 17

Captions: Horace working on his farm, see *Epistles* 1, 14, 39: 'rīdent vīcīnī glaebās et saxa moventem' (the neighbours laugh at me as I remove turf and stone).

Actium

p.172, ll.7–9: the conclusion of *Odes* 3, 16: 'Those who seek much lack much: he is well off to whom god has given just enough with sparing hand.'
p.172, para. 3 **tōta Italia . . . iūrāverat**: see *Rēs Gestae* 25, 2, in which Augustus says: 'iūrāvit in mea verba tōta Italia sponte suā, et mē bellī quō vīcī ad Actium ducem dēpoposcit' (the whole of Italy swore loyalty to me and demanded me as its leader in the war which I won at Actium).

Suetonius, *Augustus* 17, 2, also refers to this oath, which must have been taken in summer 32 BC (for the *Rēs Gestae*, see Chapter 20, Exercise 20.3, p.205).

p.173, para. 3 **cum Agrippa**: M. Vipsanius Agrippa had been serving with Octavian in Apollonia before Caesar's murder and he became, with Maecenas (whom he hated), Augustus's most trusted friend and most successful general; it was he who defeated Sextus Pompeius at Naulochus in 36 BC.

★ p.173: this relief of the first century BC from Praeneste may well date from the time of Cleopatra. Note the crocodile on the bows. The legionary soldiers stand on deck while invisible slaves work the oars below.

p.174, para 2: we have followed Plutarch's account. Modern historians have a higher estimate of Antony and Cleopatra's deliberation and courage: 'Antony was . . . forced to signal to Cleopatra, who had the war-chest aboard, to escape. He broke off the engagement and managed to join her with forty ships.' (H. H. Scullard.) Antony and Cleopatra may not in fact have wished to fight a battle but only to break out of the Ambraciot Gulf where they were being blockaded. (See C. R. R. Pelling's interesting article on this battle (*Classical Quarterly* XXXVI, 1, pp.177–81).)
p.174, para. 3, l.1: **Vergilius in Aenēidos octāvō librō**: Virgil, *Aeneid* 8, 370. Venus goes to Vulcan and asks him to make arms for Aeneas. She brings the completed products of Vulcan's workshop to Aeneas (l.608) – helmet, sword, breastplate, greaves, spear, and, above all, the shield, on which Vulcan has miraculously represented scenes from Roman history to come, starting with Romulus and Remus and ending with Actium. We quote lines 675 ff. with omissions.

The passage is not too difficult in the prose paraphrase; word order, as usual, is the main difficulty of the original.
penātibus: this normally means the spirits of the larder, in a private house; here it means the native gods of Rome, as opposed to the **magnīs dīs**, the gods common to all mankind. The rhythm of this line, with a spondee in the fifth foot and a monosyllabic ending, is very striking.
ope barbaricā: Octavian's propaganda made the war a conflict between West and East, Rome and barbarian Orient (this was a distortion of the truth; 300 senators had fled to Antony).
p.175, last para. **occāsiōnem quā . . . praebēret** an opportunity to show (relative + subjunctive, expressing purpose).

Consequence clauses

These differ from the final construction (purpose) in that (1) the negative form is **ut nōn** (2) the rules of sequence cannot apply, e.g. a past action may result in a present consequence, cf. Exercise 17.1.6; further, after an historic main verb the perfect subjunctive is regularly used if the actuality of the event is stressed, cf. Exercise 17.1.3. The tenses of the subjunctive used are in accordance with common sense and there is no need to labour the inapplicability of the rules of sequence.

Exercise 17.4

p.177, Maecenas's illness: see *Odes* 1,20, quoted below: the date of the occasion is not known; Horace refers to the incident again in *Odes* 2, 17, 26–7

> . . . cum populus frequēns
> laetum theātrīs ter crepuit sonum.

(when the people, packing the theatre, burst three times into a thunder of applause).
p.178, last para. **vīle pōtābis** . . .: *Odes* 1, 20, omitting the last verse. The invitation poem is found before Horace; in Horace's lifetime Philodemus of Gadara (fl.58 BC) wrote an epigram (*Anth. Pal.* 11, 44) inviting his Roman patron Piso to dinner, warning him that the food would be poor but the conversation good. Catullus (13) goes one better, perhaps parodying the convention, when he invites Fabullus to bring the dinner with him:

> cēnābis bene, mī Fabulle, apud mē
> paucīs, sī tibi dī favent, diēbus,
> sī tēcum attuleris bonam atque magnam
> cēnam.

(You will dine well with me, Fabullus, in a few days, if the gods smile on you, if you bring with you a good, big dinner.)
Horace uses the convention to pay Maecenas an affectionate compliment reminding him of an occasion when he was publicly honoured. This is not to say that the poem was not the occasion of an actual invitation to dinner in Rome or at the Sabine farm.

★ p.178: the model shows the theatre inaugurated by Pompey in 55 BC, the first stone theatre in Rome. Behind it are the porticos, which contained trees and fountains and afforded generous exhibition space for paintings and sculptures.

p.179, l.1 **cāre Maecēnās eques** dear knight Maecenas. Maecenas, despite the important functions he performed for Augustus, never took office and remained a knight.
(Some editors prefer the ancient correction **clāre** for **cāre**; this gives more point to **eques**: though Maecenas is only a knight, he is famous.)
ut redderet: ut expresses consequence: 'so that the banks . . . returned'. The verb is singular agreeing with the nearest subject.
The last verse (omitted) contains a textual crux and its interpretation is disputed; it probably means: 'although you drink Caecuban and grapes tamed in the Calenian press (at home), neither Falernian vines nor the vineyards of the hills of Formiae temper my cups,' i.e. you won't get any of the grandest wines here.
Question 4(b): he implies his gratitude for Maecenas's generosity by the oblique reference to the Sabine farm, the wines of which he will provide at dinner.

Exercise 17.5

This is a revision exercise; so many constructions have been introduced in the last few chapters, that it is important that pupils should be able to identify and distinguish them.

Background

Philip Francis's eighteenth-century translation is in rhyming couplets. What difference does rhyme make to the poem? David West – who has kindly written this translation especially for us – writes in free verse. Might he just have well have written in prose?

★ p.180 (and cover of Part II): the girl picking flowers, perhaps a personification of the season of spring, is from a wall-painting in a bedroom in the Villa of Ariadne near Pompeii. She floats against her background with a miraculous lightness.

Chapter 18

Bellum Alexandrīnum

The narrative content is based on Plutarch's *Antōnius*.

p.181, para. 3, l.6 **unde vidēret**: the subjunctive expresses purpose: 'from where he might see', i.e. 'so that he might see from there'.
p.182, para. 1 **eī nūntiātum est** it was announced to him: the first example of the impersonal passive, unlikely to cause difficulty, since we use the same idiom in English.
p.182, para. 2 **in altā turre**: according to Plutarch, she had shut herself in 'her monument', but he does not make it clear what he means by this.
p.182, last l. **tāle dēdecus ferre nōn potuit**: cf. Horace, *Odes* 1, 38, 25–33:

> She dared to look calmly on the ruins of her palace, she had courage to handle the cruel serpents, so that she drank their black poison in her body, all the more fierce in that she had planned her death, since she begrudged the savage Liburnian galleys the reward of having her led in Octavian's proud triumph, no humble woman.

From this we see that the tradition of the serpents goes back to soon after her death.

★ p.183 **calathus fīcīs plēnus**: this basket of figs is a wall-painting from the dining room of the villa at Oplontis near Pompeii. It was only in 1964 that systematic excavations began here. The villa may have belonged to Poppaea Sabina, the wife of Nero. She died in 65 AD. The property was unoccupied when it was buried beneath ash and pumice from Vesuvius in 79 AD.

Verbs and adjectives with ablative

p.185: the ablatives after these verbs and adjectives are not all similar to the grammarian's eye, e.g. **ūtor** is followed by an instrumental ablative, **careō** by an ablative of separation, but there is no point in introducing your pupils to such complications at this stage.

cum = when

cum + indicative: we have simplified the facts here by only introducing **cum** + indicative in future time (which is the commonest usage).

Exercise 18.5

★ p.187: this relief from Boston may be the only reliable ancient portrait of Horace. The wine-mug suggests an Epicurean and the receding hairline suggests Horace! But cf. the medal of Horace from the fourth century AD (Part III, p.194).
p.188 **quīnque diēs . . .**: *Epistles* 2, 7, 1ff.
l.2 of the verse **dēsīderor**: present of remaining effect.
ll.4–5 of the verse **quam . . . dās, dabis . . . veniam**: it is quite common in Latin for the relative clause to precede the antecedent. This can never be done in English and always causes difficulty.
l.6 **puerīs** for their children: dative of the person concerned.
l.8 **vātēs**: this word properly means a prophet. It was used of poets, the prophets or interpreters of the Muses, at least by the time of Ennius, but was later replaced by the Greek word **poēta** (= maker); the use of **vātēs** was revived to honour Virgil, and Horace uses it of himself several times. Here there is perhaps a touch of irony – 'your humble friend whom you call **vātēs**'.
Question 4 **sibi parcet**: Horace is growing older and more concerned for his health; Rome in summer brings risk of fever and in winter is too cold.
Question 6: summer, the season of excessive heat and fever, winter, the season of excessive cold and snow, when he must stay huddled up in a warm climate indoors and read. Spring brings the warm west winds and the swallow, when he can safely return to Rome.

★ p.189: this cameo of a Hellenistic queen of Egypt may represent Cleopatra. It is in the Museum of London and is part of what is known as the Cheapside hoard. This collection of treasures, the stock-in-trade of a Jacobean goldsmith living in Cheapside, London, was found in a box buried under the floor perhaps during an outbreak of plague in 1603.

Chapter 19

Caesar Augustus

p.191, l.5 **amīcum quendam**: this friend was C. Cornelius Gallus, the poet, who was also a friend of Virgil (see *Eclogue* 10). Augustus made him the first Prefect of Egypt but he fell into disgrace (for excessive ambition or vain-glory); he was recalled in 27 BC and committed suicide. Nothing of his poetry survived until in 1978 papyrus pages containing parts of several poems were recovered from a rubbish heap of a

Roman garrison town in Egypt. See *Omnibus* 1, 1–4 and the illustration on p.139 of Part II: this is part of the Gallus papyrus.
1.9 **proximō annō**: 29 BC.
para. 2 **portae templī Iānī**: the gates of the temple of Janus were closed only when there was peace throughout the Roman Empire. They had been closed only twice during previous Roman history. The Senate ordered their closure in January 29 to mark the end of the Civil War and they were closed twice more during Augustus's reign (see *Rēs Gestae* 13 and Virgil, *Aeneid* 1, 294, where Jupiter's prophecy to Venus about the future of Aeneas's descendants culminates in the prediction of the peace Augustus will bring: 'claudentur Bellī portae.')
p.191, last l. **ratī**: **ratus**, the perfect participle of **reor**, is used very commonly meaning 'thinking'.
p.192, l.2 **multī eum quasi deum colēbant**: emperor worship grew up spontaneously in the Eastern provinces, where rulers had always been worshipped, as in Egypt. Augustus regularized and encouraged the worship of **Rōma et Augustus** in the provinces as a focus of political loyalty. In Italy cults were established of which freedmen were priests (**sēvirī Augustālēs**); freeborn Roman citizens might not participate in these cults. But poets frequently refer to Augustus as a god on earth or predict his deification after death (in fact Augustus was deified by decree of the Senate immediately after his death); compare e.g. Virgil, *Georgics* 1, 503ff., Horace, *Odes* 2, 5, 1–4.
p.192, l.2 **aliī dīcēbant eum Rōmulum esse**: cf. Virgil, *Georgics* 3, 16ff. Virgil says he will build a (metaphorical) sanctuary to commemorate Actium:

> In the middle I shall have Caesar and he shall dwell in a temple; . . . on the doors I shall represent the battle and the arms of victorious Quirinus (= Romulus).

ll.3–4 **paucī ē nōbilibus eī invidēbant**: opposition from diehard republicans culminated in the conspiracy led by the consul Murena in 23 BC.
p.192, l.5 **prīnceps senātūs**: this was the title 'traditionally pertaining to the senator foremost in rank and authority' (Syme: *Roman Revolution*, p.301). It was conferred on Augustus in 28 BC. After this, **prīnceps** was the title used to describe Augustus as leader of the whole state. It had a civilian ring to it, which Augustus, in his consistent efforts to disguise the military basis of his power and appear as the bringer of peace, preferred to the military title **imperātor**. He calls himself **prīnceps** in the *Rēs Gestae*.
l.6: the **cognōmen** Augustus was conferred on him by the Senate on 16 January 27 BC, who preferred it to the alternative Romulus (Suetonius, *Dīvus Augustus* 7, 2). The word until then had only been used in religious contexts, often linked with **sānctus**.
p.192, para. 2, l.1 **triumphum triplicem**: held on 13–15 August 29 BC, to celebrate victories gained in Illyricum, Actium and Alexandria (see Background section).

★ p.192: this silver cup of the first part of the first century AD is from Boscoreale. It shows Augustus's successor, the Emperor Tiberius, riding in his triumphal chariot crowned by Victory. His lictors go before him.

p.193, l.2 **dīxit sē velle rempūblicam restituere**: we tell the story very much from Augustus's point of view. On 13 January 27 BC he announced in the Senate that he resigned all powers and provinces to the free disposal of the Senate and people. The Senate begged him not to abandon the republic and he responded by accepting for ten years an unusually large province (Spain, Gaul, Syria) with proconsular **imperium**. This gave him command of most, but by no means all, the legions. At home his power rested for the next four years on repeated consulships. He could quote republican precedents for both these arrangements. This settlement remained in force until 23 BC, when the crisis occasioned by the conspiracy of Murena persuaded him to make a new arrangement.

The **testāmentum** referred to is the *Rēs Gestae* (see Chapter 20) and we quote part of Chapter 34, which continues: 'From this time I exceeded all in influence (**auctōritās**) but had no more power than others who were my colleagues in the several magistracies.' Tacitus (*Annals* 1, 2) on the other hand dates the beginning of his **dominātiō** from this settlement.

Clauses of fearing

We say that verbs of fearing are followed by **nē** or **nē nōn**. Instead of **nē nōn, ut** is sometimes used: cf. Caesar, *B.G.* 1, 39: 'sē . . . rem frūmentāriam, ut satis commodē supportārī posset, timēre dīcēbant' (they said that they were afraid the corn supply could not be adequately provided). We omit this complication.

Pupils may have some difficulty in recognizing that whereas Latin uses only one form of words in clauses of fearing, English has several alternatives, which do not appear to differ in meaning, e.g. **timeō nē in perīculum veniāmus** = **1** I am afraid lest we get into danger. **2** I am afraid that we will get into danger. **3** I am afraid that we may get into danger.

Exercise 19.3

p.195: the first sentence exemplifies the important principles illustrated in this exercise; (1) **ille** is often used at the beginning of a sentence to introduce a change of subject (2) **sē** (and **suus**) refer to the subject of the leading verb in all forms of indirect speech. Latin is much clearer than English in its use of such pronouns.

Exercise 19.5

p.195, Horace's lowe poems: Odes 1-3 include seventeen poems in which 'love' themes are treated; none of them, in our view, is to be taken seriously, at least not wholly seriously. Horace tells us this himself in *Odes* 1, 6, 16–20:

> nōs convīvia, nōs proelia virginum
> sectīs in iuvenēs unguibus ācrium
> cantāmus vacuī, sīve quid ūrimur,
> nōn praeter solitum levēs.

(I sing of dinner parties, I sing of battles waged by girls fiercely attacking the boys with sharpened finger nails, heart-whole, or, if I do burn at all with passion, light (in tone) as usual.)

In this ode he is refusing to write of Agrippa's military exploits and says that he will keep up the traditional themes of lyric poetry – sympotic and erotic. His 'love' poems often make no pretence of being autobiographical; he frequently assumes the role of the spectator watching the battle of love with amused detachment. When he does speak in the first person, he is often laughing at himself: **simulat sē sēria dīcere sed rē vērā iocātur**, as in *Odes* 1, 22 (**integer vītae** . . .).

p.196, **vīxī puellīs:** *Odes* 3, 26.

l.1 **vīxī**: the poem opens with what is ostensibly a solemn pronouncement; his life as a lover is over (compare Dido's last words, *Aeneid* 4, 653:

> vīxī, et quem dederat cursum Fortūna perēgī.

(my life is over; I have completed the course which Fortune had given me)), but this solemnity seems to be belied immediately by **puellīs idōneus**: Lewis and Short say that **idōneus** means 'suitable' etc. 'especially for some action'; it is common in military contexts. The word suggests, perhaps, that Horace was physically fit, an able-bodied lover.

l.2 **mīlitāvī**: for love as a battle in the service of Cupid, compare Ovid (*Amōrēs* 1,9), who develops the conceit in forty-six lines:

> mīlitat omnis amāns, et habet sua castra Cupīdō

(every lover is a soldier, and Cupid has his camp)

ll.3–6: the occasion of the poem now becomes clear; on resigning from the lists of love Horace is dedicating his weapons to Venus, just as a solider on retirement would dedicate his weapons to Mars.

l.7: in the middle of the line there is a sudden break in the sense and change in the tone; **hīc, hīc pōnite** shows that he is talking to companions and the repeated **hīc** demands urgent action. What began by looking like a dedicatory epigram has developed into a dramatic monologue; that is to say, although Horace alone speaks, a drama with other actors is suggested.

The list of weapons which his companions are carrying forms a crescendo of absurdity. He himself has hung up the lyre which he used for serenading; his companions carry the shining torches which lit his way and crow-bars and bows, which 'threatened the doors which shut me out'. The theme of the excluded lover, perishing on the doorstep of his beloved, appealing to or threatening the doorkeeper, or even the door itself, was traditional and appears in Horace, *Odes* 1,25 and Ovid, *Amores* 1,6 etc. Horace is here parodying a traditional theme.

As the weapons are dedicated, Horace makes his prayer to Venus, starting with the traditional invocation, listing cult centres; but the prayer ends with another unexpected twist – that Venus should raise her whip and touch Chloe for her arrogance (in refusing Horace) and send her running back to him. Horace has not finally retired after all.

The drama could, with some speculation, be recreated as follows: Horace has gone to serenade Chloe, who has never refused him; this time she would not open her door. In high dudgeon he rushes off to the nearest temple of Venus, where he solemnly declares that his life as a lover is over and dedicates his weapons; but at the very last moment he has a change of heart.

In twelve lines Horace has developed a theme from epigram with extreme sophistication. The tone changes three times – the opening is solemn and rather pompous; the central section is urgent (note the enjambement) and, we think, ludicrous. The prayer begins with a leisurely invocation, two whole lines devoted to the cult centres, and ends on a note of sad wistfulness, the point being reserved for the last two words. The whole poem is permeated by irony; if you choose to think that it has an autobiographical element, Horace is laughing at himself.

Background

An exhilarating recreation of a Roman triumph – Renaissance in period but classical in spirit – is Mantegna's *Triumph of Caesar* in the Orangery at Hampton Court.
PS (p.198): for Augustus's triple triumph, see *Aeneid* 8, 714–28:

> But Caesar, entering the walls of Rome in a triple triumph, solemnized his immortal vow to the gods of Italy, three hundred very great shrines over the whole city.

Augustus started his programme of building and restoration in 28 BC. He says that he restored eighty-two temples.

Chapter 20

★ p.199: in this noble statue of Augustus from Prima Porta near Rome (now in the Vatican), the Cupid on a dolphin, supporting the leg, is a reference to Venus, the legendary ancestress of the Julian family and thus of Augustus.

Pax et princeps

p.199, paras. 1 and 2: it is a paradox that although Augustus represented himself again and again as the bringer of Peace, he in fact conquered more territory than any leader before him (mostly through his generals, not in person). But the purpose of these campaigns was to bring the empire up to stable frontiers; they were wars to end war. So Virgil (*Aeneid* 6, 851–3) ends Anchises's prophecy of Rome's future:

tū regere imperiō populōs, Rōmāne, mementō
(hae tibi erunt artēs), pācīque impōnere mōrem,
parcere subiectīs et dēbellāre superbōs.

(Romans, study to rule the peoples by your empire (these shall be your arts) and to make peace the norm, to spare the conquered and fight the proud into submission.)
The map on p.6 should be used to clarify these paragraphs.

p.200, l.3 **caelō tonantem . . .**: Horace, *Odes* 3, 5, 1–4.
p.200, para. 2, ll.2–3 **foedus cum Parthīs fēcit**: in 20 BC Augustus negotiated a settlement with the Parthians by which they returned the standards captured at Carrhae, where Crassus was defeated in 53 BC.
p.200, para. 2, l.5: the disaster of Varus. Augustus had intended to advance the frontier of the empire from the Rhine to the Elbe. In AD 9 Varus led three legions into the Teutoberg forest where he was trapped in an ambush by the German leader, Arminius; all three legions were annihilated. Augustus then abandoned the policy of advancing the frontier to the Elbe, and the army was reduced from twenty eight to twenty five legions.
p.200, para. 3: Virgil had completed the Georgics in 29 BC.
p.200, para. 3, ll.9–10 **ita tractābat ut . . .** he treated the story so that he brought in the future in a marvellous way. There are three passages in the *Aeneid* in which Virgil explicitly works in the future history of Rome: (1) Jupiter's prophecy to Venus (*Aeneid* 1, 257–96 – see below); (2) Anchises's review of the souls of Roman heroes waiting in the Underworld to go to the world above (*Aeneid* 6, 630–728); (3) the shield Vulcan made for Aeneas (see Chapter 17, *Aeneid* 8, 630–728). Each passage culminates in praise of Augustus. In *Aeneid* 8, 337–69, Aeneas is taken on a tour round the future site of Rome.

p.200, para. 4: we quote from *Aeneid* 1, 281 & 286ff. in l.3: **rērum dominōs gentemque togātam** ('lords of the world and the togaed race' – the Romans in their military and civilian roles); and in l.4: **Augustus Caesar, ab Aenēae stirpe ortus**: the Julian gens, into which Augustus had been adopted, claimed descent from Iulus, Aeneas' son.
l.5 **Oceanō**: i.e. the Atlantic Ocean (contrasted with the sea (**mare**) = the Mediterranean).

p.200, para. 5: *Odes* 1–3 were published in 23 BC. In the first Ode of Book 1 Horace states his ambition – to be numbered amongst the 'lyricīs vātibus.' In the last Ode of Book 2 he claims that this ambition has been fulfilled:

> dīcar . . .
> ex humilī potēns,
> prīnceps Aeolium carmen ad Italōs
> dēdūxisse modōs.

(It shall be said of me that . . . grown powerful from humble origins, I first brought Aeolian song (i.e. the lyric poetry of Alcaeus and Sappho) to Italian metres.) (The meaning of this claim is clear enough – that he was the first (and in fact the last as well) to naturalize Greek lyric metres in Latin verse; but it is oddly put, since the metres were Greek, not Italian; commentators, e.g. Quinn, suggest that he refers to the refinement of these metres in his handling of them.)

p.201, para. 1: Augustus had arranged to meet Virgil in Athens as he returned from completing the Parthian settlement. Virgil fell ill in Greece and was brought back to Brundisium, where he died on 20 September 19 BC.

The *Aeneid* was unfinished and on his death-bed he gave intructions that it should be destroyed, but Augustus forbade this and entrusted Varius and Tucca with the task of editing it for publication.

p.201, 2nd para. **nāvis, Vergilium** . . .: *Odes* 1, 3. This poem cannot in fact be associated with Virgil's last voyage to Greece since *Odes* 1-3 were published in 23 BC. **reddās incolumem precor** I pray that you may deliver Virgil safe.
p.201, 3rd para. Horace and Augustus: this passage is based on Suetonius's *Vīta Horātiī*, which contains a section on the relations between Horace and Augustus. Suetonius was Secretary **ab epistolīs** (i.e. in charge of the whole official correspondence of the Empire) under Hadrian and would presumably have had access to the letters of Augustus which he quotes. Augustus had a high opinion of Horace's poetry; Suetonius says: 'He thought so highly of his writings and was so convinced that they would last for ever, that he instructed him to compose the *Carmen Saeculāre* . . .'. When Horace refused to become his secretary on the grounds of ill health, Augustus did not resent it (Suetonius, *Vīta*): 'and not even when he refused did Augustus become angry with him or cease to press his friendship on him. There are letters extant, from which I quote a little below for proof of this: "Assume some rights with me, as if you lived in my house; you will do so rightly, with no presumption, since I wanted to have this relationship with you, if your health had allowed it." '
p.202, para. 1 the Secular Games: these were first celebrated in 348 BC, then in 249 and 146. They were held to commemorate the end of one **saeculum** and the beginning of another. The date for the celebration in 17 BC was arrived at by allowing 110 years for a **saeculum** instead of 100. The celebrations continued for three successive days and nights with sacrifice and prayer and games. Horace's *Carmen Saeculāre* was performed on the last night by a choir of twenty-seven boys and twenty-seven girls, first on the Palatine and then on the Capitol.

★ p.202: the breastplate of the Prima Porta Augustus (p.199). The imagery here has strong links with Horace's *Carmen Saeculāre*. At the lower left and right are Apollo and Diana while at the centre the recovery from the Parthians of the standards lost at Carrhae is portrayed.

p.203: Maecenas's last request to Augustus was 'Horātī Flaccī ut meī estō memor' (Suetonius, *Vīta:* 'Remember Horatius Flaccus like myself.')

★ p.205: a wall from the Temple of Rome and Augustus at Ankara. Here is the best preserved copy of the *Rēs Gestae*.

Exercise 20.3

Extracts from the *Rēs Gestae*.
p.206, Question 1 (*Rēs Gestae* 2): he refers to Brutus and Cassius and the double battle of Philippi. **in exsilium expulī** is not true; Brutus and Cassius left Italy in summer 44 BC, when

Antony was menacing them. They left ostensibly to take up their provinces of Crete and Cyrene, though in fact they went to Macedonia and Syria respectively. **cum bellum īnferrent reīpūblicae**: they would have said that they were defending the Republic and were being attacked by a faction.
Question 2 (*Rēs Gestae* 3): pupils should be able to comment on **bella cīvīlia** but will have only a general inkling of **bella externa**; the map on page 6 would help them with this. **omnibus veniam petentibus cīvibus pepercī**: he refers to the amnesty of 39 BC (Treaty of Misenum), and his much advertised **clēmentia** after the Bellum Alexandrinum. Your pupils will recall the particular cases of Horace himself, Pompeius, and, above all, Cicero **fīlius**, whom Augustus promoted to the **cōnsul suffectus** in 30 BC. But to say that those who asked for pardon received it begs the question of what happened to those who didn't. Augustus either eliminated his enemies or turned them into his clients. Thus he became the patron of whole state.
externās gentēs . . .: this is a vague statement (compare Virgil, *Aeneid* 853 – Rome's mission is **parcere subiectīs et dēbellāre superbōs**). It is certainly true that, in accordance with long Roman custom, he incorporated the territories which he conquered into the Roman Empire as provinces and did not destroy them.
Question 3 (*Rēs Gestae* 13): the doors of the temple of Janus (see Teacher's Book, p.56) were closed in January 29 BC to mark the end of the Civil War and in 25 BC to mark the successful conclusion of Augustus's campaigns in Spain. The third occasion is unknown.
Question 4 (*Rēs Gestae* 26): this question may be answered with the help of the map on page 6. Egypt was not made a province but became the personal possession of Augustus, governed by a praefectus. On the title Augustus, see Teacher's Book, p.56.
Question 5 (*Rēs Gestae* 34): discussed in Teacher's Book, p.56. If you discussed this settlement with your pupils they may see that Augustus's description of his position is slanted.
Question 6 (*Rēs Gestae* 34, 3): he here contrasts his legal power (**potestās**) with his influence (**auctōritās**). Although his colleagues as consul enjoyed formal equality with him, the claim will not bear scrutiny, especially in the light of the settlement of 23 BC, by which he was granted proconsular **imperium** throughout the empire.

Exercise 20.4

p.207 **diffūgēre nivēs** . . .: *Odes* 4, 7. In Part I, Chapter 18, we make Horace attempt to write a version of this poem during a boring lesson from Orbilius. He got as far as:

diffūgērunt nivēs, redeunt iam grāmina campīs
arboribusque comae

before he was caught by Orbilius, who gave him six of the best for inattention and for writing bad verses (**diffūgērunt**, as Orbilius pointed out, will not scan). We make him complete the poem in a context which is entirely imaginary but perhaps not unsuited to the tone of resigned melancholy which pervades the whole.

The first six lines describe the arrival of spring in three images from nature, followed by an image from the realm of myth. It is in these last two lines, in our view, that the emotion is strongest and the tone clearest, but this may not seem so to your pupils, who will be unfamiliar with the connotations of the Graces and their dance, which represent all that is beautiful and joyful.

The transience of spring and the cycle of the seasons have a message for men which is placed emphatically at the beginning of the sentence. Juxtaposed with the dance of the Graces, these words carry great weight, which is pointed by the jarring rhythm and sound. The peculiarity of the rhythm is not just that it is unusually heavy for this poem (spondees in first and third foot) but that verse stress and word stress coincide – **ímmōr|tálĭă|nē** . . . which is an effect the Augustan poets nearly always avoid in dactylic hexameters (before the end of the line); and the repeated long **ē**s of **nē spērēs** are ugly.

Structurally, the hinge of the poem comes in ll.13–14; l.13 looks back to the recurring cycle of the seasons, l.14, introduced by the emphatic **nōs**, looks forward to the one-way road which all men must take down to death.
l.16 **pulvis et umbra sumus**: the body to dust, the spirit to a shadow, both void and insubstantial.
We omit four lines which contain the Epicurean message 'carpe diem', which occurs several times in the Odes:

quis scit an ādiciant hodiernae crāstina summae

tempora dī superī?
cuncta manūs avidās fugient hērēdis, amīcō
quae dederīs animō.

(Who knows whether the gods above will add tomorrow's time to today's total? All that you have given to your own dear self will escape the greedy hands of your heir.)
In the last lines we quote, **nōn . . . nōn tē . . . nōn tē** sound out a ring of doom, the rhetorical climax of the poem.

We have omitted the last four lines, which bring the poem to a quieter close, softening the harsh message with mythical examples which show that even the favourites of the gods cannot escape Death.

īnfernīs neque enim tenebrīs Diāna pudīcum
līberat Hippolytum,
neque Lēthaea valet Thēseus abrumpere cārō
vincula Pērithoō.

(For Diana cannot free Hippolytus from the darkness of the Underworld despite his chastity, nor can Theseus break the bonds of Lethe which bind his beloved Perithous.)

The thought of the poem is crystal clear, the diction simple, the imagery unremarkable but effective (often conveyed by a single word, e.g. **rapit, prōterit, effūderit, iners**). The whole carries conviction by its restraint and achieves its effect partly by the rhythm. The metre consists of a dactylic hexameter followed by half a hexameter (up to the third foot caesura); the shorter of the two lines is always in pure dactyls and the couplets are all end-stopped except at l. 10, where the race of the seasons is pointed by enjambement between three successive lines. The regular swell and fall of the couplets, moving fast (a high proportion of dactyls), gives a feeling of urgency; the speed of the rhythm contrasts with the sombre tone of the sense; the broken hexameter cuts short the cycle and suggests the impermanence of human life. All ancient poetry was composed to be read aloud and the magic of this poem will be lost if it is read silently and not heard.

A. E. Housman surprised his class with a rare outburst of emotion after he had dissected this ode with 'the usual display of brilliance, wit, and sarcasm . . . He read the ode aloud, first in Latin and then in an English translation of his own. "That," he said hurriedly, almost like a man betraying a secret, "I regard as the most beautiful poem in ancient literature".' We append his translation:

The snows are fled away, leaves on the shaws
And grasses in the mead renew their birth,
The river to the river-bed withdraws,
And altered is the fashion of the earth.

The Nymphs and Graces three put off their fear
And unapparelled in the woodland play.
The swift hour and the brief prime of the year
Say to the soul, *Thou wast not born for aye.*

Thaw follows frost; hard on the heel of spring
Treads summer sure to die, for hard on hers
Comes autumn, with his apples scattering;
Then back to wintertide, when nothing stirs.

But oh, whate'er the sky-led seasons mar,
Moon upon moon rebuilds it with her beams:
Come *we* where Tullus and where Ancus are,
And good Aeneas, we are dust and dreams.

Torquatus, if the gods in heaven shall add
The morrow to the day, what tongue has told?
Feast then thy heart, for what thy heart has had
The fingers of no heir will ever hold.

When thou descendest once the shades among,
The stern assize and equal judgement o'er,
Not thy long lineage nor thy golden tongue,
No, nor thy righteousness, shall friend thee more.

Night holds Hippolytus the pure of stain,
Diana steads him nothing, he must stay;
And Theseus leaves Pirithous in the chain
The love of comrades cannot take away.

★ p.207: this wall-painting of the Graces, presumably based on a Hellenistic sculpture, is one of many examples of this grouping from antiquity. The Graces are Aglaia (representing splendour), Thalia (abundance), and Euphrosune (jollity). Compare Botticelli's *Prima Vera*.

Background

p.209, para. 2: for the *Georgics* passage here paraphrased, see note on p.43.
p.209, para. 4: Horace wrote:

ego nec tumultum
nec morī per vim metuam tenente
Caesare terrās. (*Odes* 3, 14, 14ff.)

(with Caesar guarding the lands, I shall fear neither riot nor violent death.)
The Ovid quotation is *Fastī* 1, 705–22. The **Ara Pācis** was set up in the Campus Martius on the banks of the Tiber in honour of Augustus's return from Spain and Gaul in 13 BC. It was dedicated on 30 January 9 BC. There was a precinct wall around the altar; the two long sides showed the procession of the Senate, the people of Rome, the magistrates, and the family of Augustus. The two short sides had four panels with mythological themes, stressing the piety and fertility of Italy. Beneath the Augustan and mythological scenes was a spreading floral pattern.
The monument was reconstructed under the inspiration of Mussolini near its original location in 1937–8. The Tacitus quotations are from *Annals* 1, 2 & 4.

★ p.211: this large sardonyx cameo, the Gemma Augustea, shows the deified Augustus seated next to the goddess Rome. (AD 10–11: Kunsthistorisches Museum, Vienna)

PART III

Part III follows a format similar to that of Parts I and II except that (1) instead of marginal glosses, notes on the text are given on the facing page; (2) there are no separate background sections; (3) there is an introduction to all six authors in the Pupils' Introduction. New linguistic features are introduced in most chapters, until the GCSE syllabus has been completed. The sequence of syntax is set out below:

THE LINGUISTIC CONTENT

Contents and sequence of grammar and syntax

Caesar

Chapter 1

The young Caesar.
Gerunds.
Compounds of **dō**.

Chapter 2

The first invasion of Britain.
Gerundives.
Compounds of **sum**.

Chapter 3

The second invasion of Britain.
Gerundives of obligation.
Compounds of **faciō**.

Chapter 4

The revolt in Gaul.
Impersonal verbs; verbs used impersonally in the passive.
Compounds of **capiō**.

Cicero

Chapter 5

The young Cicero.
Some common case usages: ablative of description, ablative absolute, partitive genitive, predicative dative.
Compounds of **eō**.

Chapter 6

Cicero's consulship and exile.
quis (anyone).
Relative with subjunctive: purpose.
Compounds of **ferō.**

Chapter 7

Cilicia and civil war.
Conditional clauses.
Compounds of **iaciō.**

Chapter 8

Cicero – the last years.
Compounds of **speciō**.

Catullus

Chapter 9

Catullus and his friends.
Subjunctive in main clauses.
Compounds of **mittō.**

Chapter 10

Catullus in love.
Passive (deponent) imperatives.
Price and value.
Compounds of **cēdō.**

Chapter 11

Catullus – the sequel.
Dative verbs revised.
Dative verbs used in the passive voice.
Dative of the person concerned.
Compounds of **pōnō.**

Virgil

Chapter 12

Aeneas arrives at Carthage.
Relative with subjunctive: consecutive and generic.
Compounds of **fugiō.**

Chapter 13

Dido and Aeneas.
Compounds of **dūcō.**

Chapter 14

The death of Dido.
Compounds of **sistō.**

Livy

Chapter 15

The greatest war in history.
Historic infinitive.
Comparative clauses.
Compounds of **currō**.

Chapter 16

Hannibal reaches the Alps.
quisque, quis, quisquam.

Chapter 17

Hannibal crosses the Alps.
Indicative and subjunctive in subordinate clauses (temporal, causal, concessive, indirect speech).

Ovid

Chapter 18

Ovid tells the story of his life.
cum and **dum.**
Compounds of **gradior.**

Chapter 19

Ovid the lover.
Numerals: distributive and adverbial.
Prolative infinitive.
Compounds of **sequor.**

Chapter 20

Ovid in exile.
Verbs and adjectives followed by the genitive.

Brief summary of Latin syntax.

THE EXERCISES

The exercises follow a pattern similar to that of the earlier Parts. All new linguistic features are practised in sentences from Latin into English, but we do not usually have sentences from English into Latin on those items 'which will be tested only in translation from Latin into English'. We do, however, provide as the final exercises of each chapter some English sentences and a short passage of continuous English for translation into Latin, to provide practice for those pupils who are taking the 'composition' option at GCSE.

We continue to give vocabulary lists of words to be learnt for each chapter, but we reduce the length of these lists in the latter parts of the book, since continuous revision of earlier vocabulary is necessary. In every chapter but four we give lists of compounds of common verbs and in Chapter 10 we give a rehash of the exercise (Part II, Chapter 20) on the formation of nouns from verbs.

In every chapter (except Chapter 16) one question is a passage of Latin continuing the extracts from the relevant author. In the case of prose authors, part of this passage is to be translated and on the other part comprehension questions are asked. In these passages both the context and the vocabulary are known; 'difficult' words or phrases are glossed. In our view this is a fairer test of the ability to translate 'unseen' than the traditional pig-in-a-poke. The comprehension questions test understanding at a sense level through straightforward factual questions; there are also simple questions on the language, and questions which require a more general understanding of the passage and some element of judgement.

On the verse authors we set similar exercises, in the case of Catullus, complete poems. We do not ask for a translation of these verse passages, nor are we prepared to recommend dogmatically to teachers how they should be handled. By the time they read the end of Chapter 9 a good class might be able to translate and answer the questions on Catullus 13 (**Cēnābis bene . . .**) without any help apart from the glosses. But only the teacher can judge whether it would be better to treat the poem orally in class before asking for answers to the questions. We would recommend that the minimum help which should normally be given is a reading aloud in Latin. The questions cover a fair range of difficulty. The first questions on each poem are at a simple sense level but some of the later questions are more sophisticated; some are looking for a personal response, although the wording of the questions is less naïve than in Part II.

We discuss at some length in our introduction to Catullus how the poems might be handled in class (see Teacher's Book, pp.76–7).

The amount of authentic Latin in Part III is likely to appear rather formidable. We hope that the whole course is so constructed that it will be possible to continue to read the Latin with some fluency and speed. But we also know from practical experience that there comes a point even in the best constructed courses when the pace falters (we anticipate that this is most likely to happen in the Virgil and Livy) and teachers may despair of getting through it all. We recommend that if this happens, the teacher should speed things along by translating parts of the extracts to their class rather than omitting any altogether. In particular, it would be a mistake to omit any complete chapter, since the successive extracts interlock and are intended to build up a solid, though incomplete, picture of the authors' work. Moreover, if any chapter is completely omitted, there will be gaps in linguistic knowledge.

It is perhaps worth saying that there is more than one way of studying a text; it is possible to read fast and rather superficially (this might be suitable for part of Caesar's swiftly moving narrative) and it is possible to read slowly and thoroughly at the greatest depth appropriate to your class (this might be the right way to treat most of the Catullus poems). We practise both methods in our reading in our own language and, if the first alternative sounds like an invitation to sloppiness, this tendency will be corrected when pupils tackle the translation and comprehension exercise on the relevant chapter.

The pupils' notes in Part III are fuller than those in the earlier parts. In the pupils' notes we have used the following conventions: **1** Where the Latin quotation is followed by no punctuation, we give a translation. **2** Where the Latin is followed by a colon, we give a comment, not a translation. **3** If a literal translation is followed by a freer translation, we use the sign =. **4** 'i.e.' is used after a translation, if we give an explanation of the meaning rather than a freer translation. **5** Where we have given the author's Latin words in a different order to assist translation, we have used italics.

The teachers' notes which follow are mainly limited to filling in the historical background of the extracts from prose authors and to making some critical comments on the poetry, which may help in class discussion of the poems. We have not aimed at giving a full and scholarly commentary but hope that what we have said may provide answers to the sort of questions which might be asked by intelligent pupils at this level. Teachers may wish to expand these notes or pass them over or to put forward alternative interpretations. Some of the notes inevitably repeat information given in notes on Part II, but we felt that we could not assume that all our readers had used the earlier parts of the course.

COMMENTARIES ON EACH CHAPTER

★Cover: a mosaic from Pompeii of the first century AD.

★Title page: portrait of Virgil from the Codex Romanus, Vatican Library, Rome. See note on ★III, p.130 (left). Note the box which contains the scrolls of his poetry and the lectern from which he recites it.

Chapters 1–4 Caesar

The best and fullest life of Caesar is that by M. Gelzer, Blackwell; that of Michael Grant: *Julius Caesar*, Weidenfeld and Nicolson, is very readable. The standard book on Caesar's Gallic campaigns is T. Rice Holmes: *Caesar's Conquest of Gaul*, Macmillan. *The Oxford History of England, Roman Britain* (Salway), Chapter 2, gives a paraphrase of Caesar's own account of his invasions of Britain with a useful commentary.

The selections from Caesar are taken from the *Dē Bellō Gallicō* (referred to as *B.G.*), Books 4 and 5, but the first chapter gives an outline of Caesar's life up to 55 BC based closely on parts of Suetonius, *Dīvus Iūlius*. This outline is not intended to be even a skeletal political biography but aims at illustrating some aspects of his character. It also forms a bridge between the made-up Latin of Part II and the authentic Caesar which follows. We wished, for instance, to introduce gerunds, very common in Caesar, in a context over which we had control.

The narrative from the *B.G.* begins at Exercise 4 of Chapter 1 with Caesar's preparations for the first invasion of Britain. We have chosen these invasions because the first contacts between the Romans and our ancestors, the 'ultimōs Britannōs' of Catullus (14, l.11), are, we felt, likely to have an immediate interest to English-speaking students. They also provide an exciting story of manageable length. In Chapter 4 we move straight on to the revolt of the Eburones and

Nervii, which sees Quintus Cicero at the centre of events; this gives a link with Part II, in which Marcus Cicero was an important character.

★ p.9: this bust of Caesar is life-size and was probably sculpted shortly before he was assassinated. It is in the Vatican.

Chapter 1

The young Caesar

p.10: Caesar made the funeral of his aunt Julia the occasion of an affirmation of his political position as a **populāris**. Julia, who was the widow of Marius, died when Caesar was quaestor, probably 68 BC. At the funeral the **imāginēs** (death-masks) of Marius's family were displayed. Caesar's first wife, Cornelia, was the daughter of Cinna, Marius's successor as leader of the **populārēs**. Sulla had intended to put Caesar to death in the proscriptions and, though persuaded not to, he remarked that any intelligent man could see that 'in this boy there were many Mariuses'.

The **gēns Iūlia** claimed Venus as their founder. The head of Venus appears on Caesar's coins and in 46 BC he dedicated a temple to Venus Genetrix.

l.17 **corōnā cīvicā**: he won this honour at the storming of Mytilene.

l.22 **apud Molōnem**: Molo of Rhodes also taught Cicero.

l.23 **dīcendī magister**: the first example of a gerund. It is glossed and explanation is best left until the reading of the narrative has been completed.

ll.24ff. = Suetonius, *Dīv. Iūl.* 4: Pharmacussa was about twenty miles south of Miletus, off the coast of Caria. Caria and Lycia were major strongholds of the pirates, who dominated the Eastern Mediterranean until Pompey swept them off the seas in 66 BC.

l.27 **ad argentum expediendum**: this is glossed. A Latin speaker would read **expediendum** as a gerundive, agreeing with **argentum**, but the sense and form will be the same whether it be gerund or gerundive. The ransom, according to Suetonius, was fifty talents, a very large sum, which was raised, according to Plutarch, by public subscription in Miletus.

l.36 **in Hispāniam Ulteriōrem**: in his first command, he conquered Lusitania (Portugal) and captured the capital of Galicia. These victories merited a triumph, but on his return to Italy the Senate refused to let him stand for the consulship in absence. He had to resign his command and so forfeit his triumph in order to enter Rome and stand in the election.

l.37 **Alexandrī Magnī imāgine**: Caesar was thirty-two (if he was born in 100 BC and was quaestor in 68 – both dates are controversial); Alexander was thirty-one when he got back to Babylon from his conquests in the East.

★ p.10: this bronze of Alexander on Bucephalus is only 49 centimetres high. Found in the Villa dei Papiri at Herculaneum, it dates from the second century BC. A lively, vigorous sculpture, it shows Alexander's hair flying as he lunges at his enemy.

l.41 **aedīlis**: according to Plutarch, he gave a show of 320 pairs of gladiators fighting in single combat besides theatrical performances, processions and public banquets. One result of the popularity he thus gained was that he was elected **Pontifex Maximus** in competition with distinguished rivals.

l.53 **societātem iniit**: the First Triumvirate was not legally established but was, as Suetonius says, a partnership between the three most powerful men in Rome. Pompey was being thwarted by the Senate and wanted land for his veterans and the ratification of his Eastern settlement. Crassus wanted, on behalf of the **equitēs**, a reduction in the amount the **pūblicānī** had bid for the contract for collecting the taxes in Asia.

l.56: the first bill Caesar brought before the Senate was to set up a land commission to distribute allotments to Pompey's veterans. The Senate threw it out. Caesar then took the bill straight to the Comitia (Assembly of the People). Bibulus attempted to obstruct the bill but was driven from the Forum by Caesar's thugs (veterans of Pompey); he spent the rest of the year shut up in his house, issuing notices that all Caesar's acts were illegal, since the omens were unfavourable.

ll.62–63 **per plēbem**: much of Caesar's legislation was carried in the **Concilium Plēbis**, through the agency of Vatinius, a tribune in Caesar's pay (**plēbiscīta**, measures passed by the plebs in the **Concilium Plēbis**, had the same validity as **lēgēs**, passed by the whole people in **Comitia**).

l.66: Cicero's letter to Atticus *ad Att.* 2, 19, 2–3. Cicero always attached great importance to displays of public opinion (especially when they were in support of himself).

★p.12: this statuette of a tragic actor is of ivory coloured blue with the sleeves striped blue and yellow. The actor, whose mouth can be seen through his mask and who is wearing high, stilt-like shoes, is playing the part of a woman. (Musée du Petit Palais, Paris)

l.82 **in prīmīs** . . .: the Senate had allotted as the consular provinces for 58 BC the woods and roads of Italy – a calculated insult to Caesar, whom they could not prevent being elected consul. Vatinius carried a **plēbiscītum** giving him Cisalpine Gaul and Illyricum with three legions for five years. When the governor elect of Transalpine Gaul suddenly died, Pompey proposed in the Senate that this province and one more legion should be added to Caesar's command.

We do not give any account of Caesar's campaigns in Gaul; their course can be followed on the map (p.14). Nor do we say anything about the complicated events in Rome during these years; these could be outlined with the help of the chronological table which precedes this chapter.

★p.14: this sculpture of a dead Celtic warrior is a Roman copy of one of the bronze statues which Attalus I dedicated at Pergamon in commemoration of his victories over the Gauls (239 BC). (National Archaeological Museum, Naples)

The gerund

The gerund itself does not occasion students much difficulty, since it corresponds closely to the English verbal noun ending in -ing. The English verbal noun looks identical with the present participle and it may be necessary to clear up this ambiguity in English by illustrating the adjectival (participle) and nominal (gerund) use of these parts of speech, but in fact difficulty is occasioned by this in translating from English into Latin rather than vice versa.

The teaching of the gerund is bedevilled by the preference of Latin authors for gerundive attraction. We have tried to write correct Latin in this chapter, using, with one exception, only gerunds of verbs used intransitively or in the genitive if transitive; this we believe is correct classical usage.

Exercise 1.3

The first extract from Caesar = *B.G.* 4, 20–21, with omissions. Caesar's lucid and straightforward prose should not often afford great difficulty and can usually be read quite fast. Sometimes pupils will be stumped at first reading by the longer sentences in which the main verb comes at the end, e.g. **itaque . . . reperīre poterat**.
l.4 **magnō ūsuī**: predicative dative, explained in Chapter 3.

★p.18: this wall painting from the Temple of Isis in Pompeii shows two Roman warships racing across the waters of a harbour. It dates from the first century AD.

ll.16–17 **pollicentur obsidēs dare** . . . **obtemperāre**: Caesar uses a prolative infinitive as in English instead of accusative and future infinitive.

Chapter 2

The first invasion of Britain

B.G. 22–36, with omissions

ll.1–2 **ad duās trānsportandās legiōnēs**: the gerundive is introduced without explanation. It will probably cause no difficulty and it is best, if possible, to delay explanation until the narrative has been completed.
ll.2–3 **quod nāvium longārum**: this use of the partitive genitive will strike pupils as odd. It may be compared with e.g. **aliquid vīnī**, familiar from Part II.

★p.20: this model of an Iron Age chariot, drawn by two small horses, is from the National Museum of Wales.

ll.19–20 **mīlitibus** . . . **dēsiliendum erat**: the gerundive of obligation is explained in Chapter 3. At present pass over it.
l.37 **pugnātum est**: the impersonal use of the passive is explained in Chapter 4.

★p.21: this model of a large sea-going freighter sailing square-rigged is in the collection of the Department of Classics of New York University.

★p.23: this illustration shows a skiff towing a vessel into harbour. It is a relief on a tomb in the Isola Sacra, the cemetery for the harbour of Rome.

Gerundives

Gerundives can seldom be translated literally and pupils will need to recognise the linguistic pattern and translate into appropriate English. They usually find this easier than one might expect.

Exercise 2.3.

Question **6(b)**: a cursory reading of this chapter might suggest that the Britons had been rather feeble. In fact they were not fully mobilized, as they were the following year; only the tribes of the coastal region were engaged. These put up a very stout resistance to the initial landing. They then ambushed one of Caesar's two legions when it was foraging and clearly came close to defeating it; lastly, undismayed, they made an attack on Caesar's camp. But in set battle they had little chance against Caesar's highly trained legions. The campaigns of 54 show that they had learnt from this experience.

Chapter 3

The second invasion of Britain

= *B.G.* 5, 8–23, with omissions.

ll.6–7 **admodum laudanda** extremely praiseworthy: pupils should see the meaning of this phrase.
l.23 **locum ēgregiē et nātūrā et opere mūnītum**: this was probably the fortress at Bigbury. See Collingwood: *Roman Britain,* p.44: 'There is an ancient fort, now called Bigbury, on the brow of the hill two miles west of Canterbury, overlooking the river. It had been constructed as a fortified town by the Belgae not long before Caesar's time.' The river is the Great Stour.

⋆ p.30: this **testūdō** is a relief from the Column of Marcus Aurelius, erected in imitation of Trajan's Column. It celebrated Marcus Aurelius's victories over the Germans and Sarmatians (169–76 AD).

l.29 **māne** in the morning, early: not to be confused with **manē**.
ll.34–5 **subsisterent . . . possent**: the verbs are in the subjunctive because they are in reported speech (part of what the messengers said). This need not be explained, unless questions are asked.

This paragraph illustrates the frequency with which Caesar changes from the aorist to the historic present. Pupils will soon become familiar with this, a common feature in poetry and historical narrative. After the historic present Caesar seems to use historic or primary subjunctives at whim.
ll.46ff.: Caesar's advance to the Thames brings him to the interior of Britain as opposed to the coastal region. The text of the *B.G.* here has three chapters of digression on the inhabitants of Britain and its geography (12–14); these form an awkward break in the narrative and some editors do not consider them authentic. We have included this extract because we thought it would interest pupils but you may prefer to omit the paragraph. The information about the tribes of the interior is inaccurate; they were more civilized than Caesar allows. They did grow corn and wore linen and woollen clothes.
ll.46–7 **quōs . . . dīcunt**: **quōs nātōs (esse) in īnsulā ipsī memoriā prōditum (esse) dīcunt**: a difficult clause, rather awkwardly expressed.
ll.58ff.: we have had to abbreviate these chapters considerably. The original shows how successful and skilful Cassivellaunus's guerilla tactics were and what difficulties the Romans had in combating his hit-and-run attacks.
l.88ff. **oppidum Cassivellaunī**: the town of Cassivellaunus was probably the hill fort north of St. Albans at Wheathamstead. This was a fortress about 100 acres in extent, protected by a rampart and ditch still forty feet deep in places.

⋆ p.33: the coin of the Catuvellauni portrays their king Tasciovanus. It was minted in Camulodunum between 20 BC and 10 AD.

l.106: **penderet** what Britain was to pay: an indirect deliberative subjunctive.

Gerundives of obligation

We take the construction step by step, providing analysis and literal translation, but the idiom needs to be grasped as a whole, the Latin dative of the agent being recognized as the logical subject. This is especially clear in the case of intransitive verbs – **nōbīs eundum est** can hardly be translated literally and the pattern must be seen straight off. More practice than we provide may be necessary.

Exercise 3.4

As a variation from the military narrative, we introduce three letters from Cicero. You may wish to do the first orally, to reduce the exercise to a more manageable length.

1 The letter to Quintus Cicero is *ad Q.Fr.* 2, 16, 4. Quintus Cicero wrote poetry which his elder brother considered better than his own. Later in this letter Cicero asks his brother what Caesar thinks of some poems he had sent him: 'quōmodonam, mī frāter, dē nostrīs versibus Caesar? . . . He wrote to me earlier that he had read the first book and said that he had never read anything better even in Greek than the early part.'

It is amazing that while Caesar was preparing to invade Britain he had time to read and comment on Cicero's verses; did he pay Cicero this fulsome compliment with his tongue in his cheek?

2 The letter to Caesar is *ad Fam.* 7, 5. At this time Cicero was evidently on better terms with Caesar than previously, but there seems a touch of artificiality in the way he writes, compared with the spontaneity of his letters to Atticus and his brother. 'Gaius Trebatius was a rising young **iūris cōnsultus** who was hoping to gain experience, advancement, and (not least important) money in Gaul and Britain from the activities and profits of conquest' (Stockton). No doubt he took this letter of introduction with him when he set out from Rome to join Caesar in April 54 BC. Caesar must have been inundated with such letters; he took on Trebatius, no doubt partly to please Cicero, but he was a talented man who later became a leading **iūris cōnsultus** and adviser to both Caesar and Augustus.

3 The letter to Trebatius is *ad Fam.* 7, 7.
l.5: Balbus, a native of Cadiz, had become a Roman citizen in the late 70s when he was serving under Pompey in Spain. He served under Caesar when he was governor of Further Spain and again in Gaul; he later became Caesar's confidential agent in Rome. After Caesar's death he joined Octavian and was appointed consul in 40 BC, the first provincial to become consul.
l.7 **quod volumus**: i.e. profit and advancement in Caesar's service.
l.12 **nē ipse tibi dēfuisse videāris**: Trebatius was suffering from homesickness (**dēsīderia urbis et urbānitātis**), as is clear from an earlier letter Cicero wrote him. He did not want to go to Britain, fearing the sea crossing.

Chapter 4

The revolt in Gaul

= *B.G.* 5, 26–52 (with substantial cuts and adaptation of Chapters 26–39)

ll.5–6 **ā duōbus prīncipibus Ambiorige et Catuvolcō**: Ambiorix and Catuvolcus were joint kings of the Eburones. Ambiorix was the moving spirit in this rebellion and fought on longer than almost any other Gallic chieftain; he was clearly in Caesar's eyes a most dangerous man. After destroying the forces of Sabinus and Cotta, it was he who roused the neighbouring tribes, the Atuatici and the Nervii, to attack Cicero's camp. He survived defeat at Caesar's hands and continued the rebellion the following year with the Treveri etc. In 53 BC he narrowly escaped capture and, when Vercingetorix had surrendered (52 BC) and all the rest of Gaul was finally 'pacified', Caesar was still in 51 pursuing Ambiorix and was reduced to depopulating the territory of the Eburones. But Ambiorix was never caught.
l.21 **Sabīnus hīs verbīs dēceptus**: throughout this passage Caesar repeatedly condemns the incompetence and cowardice of Sabinus, contrasting his behaviour with the prudence and gallantry of Cotta. The destruction of these forces was the most serious loss Caesar suffered in his Gallic campaigns.

★ p.41: from Trajan's Column: the soldiers on the right are felling trees while those on the left are fetching water. Also on the left, Trajan points to a fort deserted by the enemy.

l.63 **oppugnārī . . . licēre**: infinitives of indirect statement (part of what the Gauls said). Pupils should recognize this.

★ p.43: from Trajan's Column: auxiliary soldiers defend the fort from the Dacian attack. The Dacian figures are portrayed in a heroic light.

l.99 **ad turrim**: if pupils notice this form of the accusative, it might be worth telling them that third declension nouns with stems in **-i** originally declined: **turris**, **turrim**, **turris**, **turrī**, **turrī**.
l.122 **hostibus**: dative of the person concerned (see Chapter 11), 'in the eyes of the enemy'.

Impersonal verbs

These should always be learned with a person attached: **mihi licet**, **mē oportet** etc.

★p.48: this inscription, carved in Caesarian capitals, was the dedication of a statue to the deified Julius Caesar (*c.* 44 BC). It means: '(This statue) was set up in honour of the deified Julius Caesar by order of the Roman people by the Lex Rufrena.' (Vatican Museum, Rome)

Exercise 4.3

ll.4–5 **cum simulātiōne agī timōris**: **timōris** is separated from **simulātiōne** for emphasis, a common type of word order with which pupils must now become familiar.
l.14 **sīc utī** so that (**utī** = **ut**, not glossed this time).

Our story ends with the relief of Cicero. You may wish to outline subsequent events – the revolt of Vercingetorix, the pacification of Gaul, and Caesar's later career.

★p.49: this relief is from the Arch of Constantine (315 AD) which was made up of fragments from older monuments. Gibbon described it as 'a melancholy proof of the decline of the arts, and a singular testimony of the meanest vanity'. However, situated next to the towering ruin of the Colosseum, it makes a powerful impression. The general is probably Hadrian.

★p.50: this poignant statue, representing a Celtic warrior who lies mortally wounded on the ground awaiting death, is a Roman copy of one of the statues dedicated by Attalus I at Pergamon. Cf. ★III, p. 14. (Capitoline Museum, Rome)

Chapters 5–8 Cicero

The following are perhaps the most useful books on Cicero:
D. R. Shackleton Bailey: *Cicero*, Duckworth.
D. L. Stockton: *Cicero, a political biography*, Routledge.
On the letters: How: *Select Letters*, Oxford, in two volumes with an exhaustive commentary; Shackleton Bailey: *Select Letters*, Cambridge; Stockton: *Thirty-five Letters of Cicero*, Oxford.

We place the extracts from Cicero in a biographical context written in simple Latin, so constructed that it sometimes anticipates the difficulties which occur in the passages quoted. We do not aim at giving a political biography; those who used Parts I and II of the course would do well to read again the background sections to Part I, Chapter 17 (on Cicero) and Part II, Chapter 5 (on Antony, Octavian and the Senate) and 7 (on the Second Triumvirate). We try to build up a picture of Cicero's character, using only letters from Chapter 6 onwards. The earliest surviving letter dates from 68 BC and so in the first chapter we use extracts from dialogues and speeches.

Chapter 5

The young Cicero

l.1 **Arpīnī**: Arpinum had enjoyed full citizenship from 188 BC. Catiline called Cicero **cīvis inquilīnus** (an immigrant citizen). None of Cicero's family had played any part in politics at Rome before him.
ll.10–27 *dē Lēgibus* 2,1, 2–3.
l.17 **tandem**: used quite commonly to emphasize interrogatives.
l.24 **scītō**: the so-called emphatic form of the imperative, commonly used from **sum** (**estō**), **caveō** (**cavētō**) and the only imperative form of **sciō** in use.

★p.53: from the House of the Wooden Shrine at Herculaneum. The shrine (at the top) takes the form of a little temple with Corinthian columns. It contains statues of the household gods. The cupboard contains glassware and ornaments.

l.41 **in rēbus forēnsibus**: 'the affairs of the forum' means the public affairs which took place there, i.e. politics, government and the law. In the law you could make your name either by pleading in the courts, as Cicero did, or as a **iūris cōnsultus**, an expert in the **iūs cīvīle**, which was a complex body of law requiring expert knowledge both for its interpretation and for getting the right legal formula for any given case. Clients consulted these experts to find out whether they had a case in law and, if so, under what legal formula.
l.41 **scīlicet**: this adverb (originally **scīre licet**) means something like 'obviously', 'of course'. It is also used ironically when the assertion in which it occurs is plainly untrue.
l.47: Servius Sulpicius Rufus, an old friend of Cicero, was one of the leading **iūris cōnsultī** of his day. When Murena defeated him in the

consular elections for 62 BC, he prosecuted Murena for electoral bribery. Murena was probably guilty but Cicero's defence secured his acquittal. In this passage Cicero ridicules Sulpicius's claim to be worthier of the consulship than Murena, a successful general; it provides a good specimen of Cicero's rhetorical style without being too difficult.
Characteristics of this style are: the use of adversative asyndeton (i.e. contrasted clauses without connection) with emphatic pronouns (**tū . . . ille**); triplets with anaphora (**haec . . . haec . . . haec**), in which the third member is the most weighty; doublets, e.g. **haec forēnsis laus et industria** (the praise won here in the forum by our efforts), **in tūtēlā ac praesidiō** (safe in the protection of).
l.48 **quī potest dubitārī quīn?**: **quī** is an old form of the ablative (how?). Verbs of doubting used negatively are followed by **quīn** + the subjunctive; the question here is equivalent to a negative.
l.74 **prō Sextō Rosciō**: Cicero defended Roscius of Ameria against a charge of parricide in 80 BC. He had in fact made his debut in the courts the year before but it was in the *Prō Rosciō* that he made his mark, boldly attacking Chrysogonus, Sulla's freedman, who placed the name of Roscius's father on the list of the proscribed in order to get his property and then accused Roscius of parricide.
ll.78–88 = *Brūtus* 306 (this work, dedicated to Brutus, was a survey of the great orators of the last century).
l.80 **accēdit labor** the toil . . . is added: **accēdit** is regularly used with a meaning equivalent to the passive of **addō**.
On this trip to Asia Cicero studied rhetoric under various masters, including Molo of Rhodes, who is said to have curbed the exuberance of his youthful style.

Common case usages

Most of these have occurred in the Latin narrative already and have been glossed so far.

Exercise 5.3

Prō Planciō 64–5: the speech was delivered in 54 BC, long after Cicero's consulship (cf. **mē . . . in maximīs imperiīs**). Vanity was one of Cicero's weaknesses but it is counterbalanced by an ability to laugh at himself.

The lesson he learnt from this experience was that the Roman people have rather dull hearing but very sharp eyes: 'I made sure after this that they should see me in person every day; I lived before their eyes and stuck close to the Forum.'

Chapter 6

Cicero's consulship and exile

l.2 **C. Verrem**: Verres's governorship of Sicily was notorious but he was confident that he would be acquitted. He was defended by the leading advocate of the day, Q. Hortensius, and the defence tried by every device to postpone the trial until the following year, when Hortensius would be consul and the presiding praetor a friend of Verres. To prevent this, Cicero made a very short opening speech (*Actiō Prīma*) and followed this immediately with the presentation of the evidence; this was so damning that Verres threw up the case and hurried off into voluntary exile. Cicero was unable to deliver the rest of his speech but published it in five books (*Actiō Secunda*).

l.14 *ad Att.* 1, 5, 2 & 8: this is the earliest surviving letter of Cicero, written November 68 BC.
l.15 **quod . . . scrībis** as for what you write . . .: this use of **quod** is common in Cicero.
l.33 **L. Iūliō. . . cōnsulibus**: Cicero uses this formula as a joke. Caesar and Figulus did not, of course, become consuls until the following January.
l.36 **Catilīnam**: Catiline was prosecuted for extortion as governor of Africa and was consequently debarred from standing for the consulship in 66 BC; in 65 he was acquitted and the following year stood for the consulship of 63 in competition with Cicero and Antonius. There were rumours that Catiline plotted to kill the consuls and seize power early in 65 (cf. Cicero, *Cat.* 1, 15) but modern scholars dismiss the so-called first Catilinarian conspiracy as propaganda, and this view is supported by the fact that Cicero thought of running in harness with him. Catiline was already notorious as a disreputable character (which is probably why Cicero gave up the plan outlined in this letter) but his revolutionary schemes seem to date from after his second rejection at the polls.
l.52 **amīca coniūrātī**: Sallust (*Catiline* 23) gives the name of the conspirator as Q. Curius and

his mistress as Fulvia.
ll.56–7 **senātū convocātō**: Sallust (*Catiline* 50–3) gives a detailed account of this famous debate. The consul designate, Silanus, called to speak first, proposed the death penalty and this proposal was supported by the next sixteen speakers. But when Julius Caesar was called to speak, as praetor designate, he said that the property of the prisoners should be confiscated and that they should be kept prisoner in the strongest municipal towns of Italy for life. Following speakers were swayed towards his proposal until Marcus Cato made a brilliant speech in favour of the death penalty. He carried the senate with him and a decree was passed in the terms he had proposed. Cicero immediately carried out the decision of the Senate; on his orders the state executioner strangled the five prisoners in the Tullianum beneath the Capitol. Cicero came out and announced to the waiting crowds: 'vīxērunt.'
l.65 **Clōdius**: we have passed over the formation of the first Triumvirate and Caesar's consulship. When Caesar went to Gaul, he needed a friendly tribune to look after his interests in Rome. He chose Publius Clodius, who belonged to the powerful patrician family of the Claudii but had adopted plebeian status so that he could become a tribune of the **plēbs**. Caesar, in order to remove a potentially dangerous opponent from Rome, had first offered Cicero a post on his staff in Gaul. When Cicero refused this, he allowed Clodius to go ahead with securing Cicero's exile. In February (58 BC) Clodius announced a proposal that anyone who had put a Roman citizen to death without trial should be forbidden fire and water.

Despite support from both senators and equites, Cicero got no help from Pompey and at the end of March he left Rome, on the very day when Clodius carried his bill and published another formally declaring Cicero an outlaw. Throughout the rest of 58 BC repeated attempts were made to recall Cicero, but all were frustrated. It was not until the following June that a series of motions were passed in Senate and assembly in favour of Cicero and finally (4 August) a law was passed by the people recalling him.
ll.73–85: *ad Fam.* 14, 2, written from Thessalonica on 4 October 58 BC: the letters Cicero wrote in exile are filled with self-pity and self-reproach.

★ p.65: this famous painting by Cesare Maccari (1840–1919) hangs in the Palazzo Madama, which has been the seat of the Italian Senate since 1871.

ll.89–103: *ad Att.* 4,1, 4–5, written from Rome in September 57 BC.

★ p.66: a page from a mid-fifteenth century manuscript of Cicero, written in formal humanist script. Note the differences between the text we have printed above and that of the manuscript. (MS. Add. c.139 Bodleian Library, Oxford)

★ p.67: this gate is not the Porta Capena, which has not survived. It is the Porta Appia (now called the Porta San Sebastiano), the most imposing gateway in the Aurelian Wall of Rome. The two towers are medieval.

Exercise 6.3

ll. 7–19: *ad Att.* 4, 3, 2–3

★ p.70: the illustration shows the ruins of Domitian's palace on the side of the Palatine Hill which rises South of the Forum to a height of 51 m.

l.17 **Tettī Damōnis**: we know nothing about this man.
l.18 **ipse**: Clodius
l.18 **diaetā**: ancient medicine recognized three branches: dietetic (not simply diet but general regimen); pharmaceutic; and surgical.

Chapter 7

Cilicia and Civil War

l.6: Clodius was aedile in 56 BC. He impeached Milo **dē vī** (for riot) before an assembly of the people (**comitia tribūta**) in the Forum.
ll.10–25 *ad Quīntum frātrem* 2, 3, 2–3, written from Rome on 13 February 56 BC.
ll.17–18 **versūs obscēnissimī**: there was said to have been an incestuous relationship between Clodius and his sister (cf. *Prō Caeliō* 32: 'quod quidem facerem vehementius, nisi intercēderent mihi inimīcitiae cum istius mulieris virō – frātrem voluī dīcere; semper hīc errō').
ll.22ff.: Cicero's style becomes more telegrammatic as the excitement rises – historic infinitives, short sentences, omission of verbs.
ll.26ff.: the gangs of Milo and Clodius met near

Bovillae on the Appian Way about twelve miles south of Rome on 17 January 52 BC, probably by chance. In the ensuing fight Clodius was killed by Milo's men. The next day a mob of Clodius's supporters burnt his body in the Forum and the flames spread to the senate house and destroyed it. Riots continued, although the Senate declared martial law. Eventually Pompey was elected sole consul and immediately carried a law **dē vī**, under which in April Milo was prosecuted. Cicero defended him; but Pompey had surrounded the court with armed guards to preserve order. Faced with this display of force, Cicero broke down; Milo was condemned by thirty eight votes to thirteen. Cicero later published the speech he had intended to deliver; it survives.

★ p.73: behind the columns of the Temple of Castor stands the Cūria. This was begun by Sulla in 80 BC and rebuilt in 44 by Julius Caesar after its burning. Marble originally covered the brick face in the lower area, with stucco above. To the left is the triple Arch of Septimius Severus (203 AD).

l.35: Pompey's law **dē prōvinciīs** enforced a five years' interval between holding office in Rome and governing a province; to fill the gap thus caused, the Senate decreed that all qualified ex-magistrates who had not yet governed a province should do so now, in order of seniority.
ll.42–3 **iniūriās prōvinciālium**: Cicero's predecessor had been Clodius's elder brother, Appius Claudius Pulcher, who, according to Cicero, had left the province 'perditam et plānē ēversam in perpetuum' (*ad Att.* 5, 16, 2).
ll.45ff.: *ad Att.* 5, 20, 2–3, written 9 December 51 BC from camp at Pindenissus. This letter will need continual clarification from the map. It is amusing to find Cicero so elated by his military success. He was very anxious to win a triumph and on his return to Italy would not enter Rome, despite the political crisis, as this would have meant forfeiting the opportunity of getting one.

★ p.76: this dramatic mosaic from Pompeii vividly conveys a ferment of movement and emotion.

l.66 **Caelius**: M. Caelius Rufus was a brilliant but unstable young man, a pupil and friend of Cicero, and a friend also of Catullus until he supplanted Catullus as Clodia's lover (see Chapter 10). Seventeen letters survive written by Caelius to Cicero while he was in Cilicia.
ll.70–85: *ad Fam.* 8, 14, 2–2, written August 50 BC.
l.73: Lucius Domitius Ahenobarbus, a leading optimate, had been beaten by Mark Antony in an election for the college of augurs.
ll.74–80: a very shrewd statement of the causes of the Civil War.
ll.79–80 **sē salvum nōn posse**: if Caesar had returned to Rome unprotected by office, he would undoubtedly have been prosecuted for offences in connection with his consulship (all his measures passed as consul were technically illegal).
l.82: since Crassus's defeat at Carrhae, another Parthian war appeared inevitable.
l.86 **mox discessūrus erat**: Cicero wrote this letter (*ad Fam.* 2,12) from camp in south east Cilicia in late June 50 BC, when his tour of duty (**annuum mūnus**) had about a month to run.
l.89 **sollicitus eram**: the so-called epistolary imperfect; so also **adferēbantur**.
l.102 **ad Tīrōnem**: *ad Fam.* 16, 11, written from Rome, 12 January 49 BC (two days after Caesar had crossed the Rubicon). For another letter written to Tiro while he was ill at Patras, see Part II, Chapter 15, Exercise 15.5.

Exercise 7.3

After crossing the Rubicon on 10 January, Caesar advanced on Rome unopposed except for a short delay at Corfinium, where Domitius Ahenobarbus, against Pompey's orders, attempted to make a stand. Pompey abandoned Rome and headed for Brundisium. Caesar took Rome and hurried in pursuit of Pompey. He wrote this letter (*ad Att.* 9, 6a) on the road between Arpi and Brundisium. Apart from his admiration and affection for Cicero, Caesar was anxious to win his support since his adherence would give Caesar's position a cachet of respectability.
l.4: Furnius, as tribune in 50 BC, had supported a public thanksgiving (**supplicātiō**) for Cicero's victories in Cilicia.
ll.7–8 **ita dē mē merēris**: the active and deponent forms of **mereō/mereor** are equally common.

Caesar reached Brundisium on 9 March and besieged it but could not prevent Pompey from getting away with his forces on 17 March. He met Cicero on 28 March but could not persuade him to support his cause. Cicero was

convinced that it was his duty to support Pompey and the cause of constitutional government; his mind was made up but he still hesitated to take the final step. Eventually he sailed from Caieta, a small port near his villa at Formiae. He wrote the following letter on 7 June on board ship (*ad Fam*. 14, 7, 2).
l.18 **nōs**: Cicero had with him the young Marcus (cf. l.26 **Cicerō bellissimus**) and perhaps also his brother Quintus and his nephew.
The letter abounds with affection for Terentia and anxiety on her behalf; it is in striking contrast with the last letter he wrote to her after his return from Greece (see p.84, l.32 ff.).

Chapter 8

Cicero – the last years

l.4 **proeliō apud Pharsālum**: the battle took place on 4 August 48 BC. In the intervening time Caesar had marched back to Spain to defeat the republican forces at Ilerda and then followed Pompey to Greece.

After the battle he pursued Pompey to Egypt, only to find that he had been murdered as he stepped ashore three day before. Cicero, on hearing of Pompey's death wrote to Atticus: 'nōn possum eius cāsum nōn dolēre; hominem enim integrum et castum et gravem cognōvī' (*ad Att*. 11, 2, 5).

Caesar had arrived in Egypt with a force of only 4,000 men; the mob rose against him and he found himself besieged in Alexandria for the winter 48–47 BC until reinforcements arrived. It was during this enforced stay that Cleopatra had herself smuggled into the palace and became his mistress.

When Caesar was eventually released from Alexandria, he had to hurry to Asia Minor to defeat Pharnaces, the son of Mithridates, in a lightning campaign (**vēnī, vīdī, vīcī**). He then at last returned to Italy, landing at Tarentum in September.

After Pharsalus Cicero returned to Italy and stayed at Brundisium until the following September (47 BC), when he went to meet Caesar at Tarentum.
l.8 **paulō post**: in fact Caesar invited himself to dinner with Cicero two years later in December 45, when he had returned to Italy after his last campaign in Spain.
ll.11–26: the letter is *ad Att*. 13, 52.
l.12 **ō hospitem . . . molestum**: we have adapted and simplified the opening sentence, which reads in our MS: 'o hospitem mihi tam gravem 'ἀμεταμέλητον'. 'ἀμεταμέλητον' means 'not to be regretted'.
l.28 Between 46 and 43 BC Cicero wrote two works on rhetoric and eleven on philosophy.
l.30 **dīvortium**: the divorce took place in January 46 BC. The reason for Cicero's estrangement from Terentia appears to be that he suspected her of extravagance and dishonesty in the management of his finances while he was away. The following December he married a young heiress who was his ward.
ll.41–47: *ad Att*. 12, 15, written on 7 March 45 BC from his remote villa at Astura, near Antium. Tullia had died in February shortly after giving birth to a son.

Servius Sulpicius's letter of condolence (*ad Fam*. 4, 5) was written from Athens in mid-March. Sulpicius, the great jurist, an old friend of Cicero, was governor of Achaea in 46–5. He died in 43 on an embassy to Antony and Cicero delivered a fine eulogy of him (*Ninth Philippic*).
l.64 **dē fīliō Mārcō**: the young Marcus, who had wanted to serve under Caesar with his cousin Quintus, agreed to go to Athens to study at the Lyceum under Cratippus. He set out from Rome in March 45. Reports soon came back from Athens that he had got into bad company and was squandering the generous allowance his father had made him.
ll.66–7: C. Trebonius met Marcus in Athens in May 44 BC on his way to govern Asia. Marcus asked if he could come and visit Trebonius in Asia; Trebonius agreed, provided he brought Cratippus with him and continued his studies.

Cicero had played no part in the conspiracy which led to the murder of Julius Caesar but two letters (one to Cassius, *ad Fam*. 16, 20) show that he approved the deed. He attended an important meeting of the Senate on 17 March, at which he proposed an amnesty and the confirmation of Caesar's acts. After this he withdrew from politics for six months and devoted himself to literature, while the political situation steadily deteriorated as Antony made his bid for power.

Cicero decided to visit Greece and see for himself how Marcus was doing. After much dithering, he finally sailed from Pompeii on 17 July, but at Syracuse he heard that there was to be a meeting of the Senate on 1 September and that there was some hope of reconciliation between Anthony and the conspirators. He therefore returned to Italy. On the way to Rome he met Brutus (17 August) from whom he heard that there had been a final rupture between Antony and the conspirators and that Antony was consolidating his power by courting the people.

Cicero did not attend the Senate on 1 September but the next day delivered his first attack on Antony in the latter's absence. After this he lay low until Antony departed for Gaul at the end of November. He then published a scathing pamphlet attacking Antony (*Second Philippic*) and on 20 December spoke against him in both the senate (*Third Philippic*) and before the people (*Fourth Philippic*). He thus placed himself at the head of the movement to resist and overthrow Antony, which led to his death the following year (9 December 43 BC).
l.88 **ad Mutinam**: at the battle of Mutina (April 43) Antony, who was investing the town, was defeated by the two consuls (Hirtius and Pansa) and Octavian. But both consuls died in the battle, leaving Octavian in control of the combined armies.

Grammar

No new syntax is introduced in this chapter; Exercises 8.1 and 8.2 are revision.

Exercise 8.3

The account of Cicero's death is a fragment of Livy, Book 120.
l.3 **Antōniō . . . Caesarī**: these are datives of the person concerned (or datives of disadvantage); this case usage is illustrated in Chapter 11.
l.5 **ut . . . cōnscēnsūrus**: intending to board: **ut** with the future participle can express purpose.
Question 10: Chapters 6 to 8 contain extracts from eleven letters of Cicero; to revise these extracts and answer this question should be a possible task for a good class.

★ p.91: the ruins of the **rōstra**, brought forward from the original site in front of the **Cūria** during Caesar's restoration in 44 BC. Columns supporting commemorative statues rose from the platform. The first structure was adorned with the **rōstra** (= beaks) of the ships captured at the Battle of Antium.

Chapters 9–11 Catullus

Books worth consulting:
K. Quinn: *Catullus, The Poems*, Macmillan (the most convenient modern edition).
James Michie: *The Poems of Catullus*, Panther Books (text with good translation on facing page).
T. P. Wiseman: *Catullus and his World*, Cambridge.

Of all ancient writers Catullus is the most sympathetic to young readers and in our selection we have chosen poems which we considered most likely to evoke a personal response from students at GCSE level. We have limited ourselves to shorter poems, most of which can, we hope, be assimilated at one go. Important aspects of Catullus's art are not illustrated; there is little evidence here, at least superficially, that he was **doctus**, and we have not used poems which would necessitate an explanation of his Alexandrianism.

We have again used a biographical framework to provide a context for the poems and have grouped them round three topics: **1** Catullus and his friends; **2** Catullus in love; **3** The sequel. We have no ancient biography of Catullus and reconstructions of his life rely heavily on the internal evidence of his poems. This evidence enables us to make a plausible reconstruction of the years when he was writing (61–55 BC?).

We do not intend to give a critical introduction to Catullus but it is worth stressing his originality. The group of poets to which he belonged, the **poētae novī**, as the poetically conservative Cicero contemptuously calls them, consciously broke with the old tradition of Roman poetry, which since its beginnings (Livius Andronicus, fl. 240 BC) had been limited to what might be called 'public poetry' – epic and drama, ultimately based on classical Greek models. The Neoterics — the name given to Catullus's group — drew their literary inspiration from the poets who flourished in Alexandria about 250 BC, above all from Callimachus, to whom Catullus implicitly acknowledges his debt in several poems.

Callimachus rejected epic as a viable genre and says that his poetry will be small scale, light, and original:

Others may bray like the long-eared ass,

But may I be the miniature, the winged poet.

(From the Prologue to the *Aetia*, in which he lays down his literary principles.)

This passage is paraphrased by Virgil (*Eclogue* 6, 1–5), echoed by Horace (*Odes* 1, 6, 5–10), acknowledged by Propertius (3, 1, 1ff.). The Augustan poets were all, in a sense, Neoterics; there was no going back from the achievement of Catullus.

Catullus prefaces his collection with a dedication to Cornelius Nepos, which begins: 'cui dōnō lepidum novum libellum?'

Here the adjectives and the diminutive describe not only the appearance of the book but also the nature of its content, which will be witty and novel. Catullus's originality has several facets. First in his choice of subject matter; he uses poetry as a vehicle for expressing his feelings on any topic which comes to mind, from the trivial to the profound. No Roman poet had done this before; indeed, before Catullus personal feelings had hardly been the subject matter of Latin poetry at all. Secondly, he was the first to use Greek lyric metres in Latin poetry (a fact which Horace chooses to ignore), and he develops the elegiac couplet, only used in Latin for epigrams until his time, in longer poems and so becomes the founder of Latin elegiac love poetry. Thirdly, his diction is strikingly different from that of his predecessors; he could handle poetic diction if he wished to (e.g. in the *Peleus and Thetis*, Poem 64), but in the shorter poems he often uses colloquial diction quite freely and with superb effect.

It is pure chance that Catullus's poems have survived. Quotations in ancient authors show that they were well known until the end of the second century; after this references to Catullus are rare (his poems were not part of the school syllabus). Transmission of the text must have become most precarious. Ultimately the survival of his poems depended upon the rediscovery of one single manuscript in the early fourteenth century by Benvenuto Campesani (d.1323); this was lost again soon afterwards but not before copies were made, of which three survive. In two of these copies the 'resurrection' of Catullus is recorded:

Versūs dominī Benevenūtī dē Campexānīs dē Vīcenciā de resurrēctiōne Catullī poētae Vērōnēnsis:

Ad patriam veniō longīs ā fīnibus exul;
  causa meī reditūs compatriōta fuit . . .
quō licet ingeniō vestrum celebrāte Catullum,
  cuius sub modiō clausa papīrus erat.

(Where did he find the MS? **sub modiō** under a corn measure; this seems poor sense, **modius** might also mean a box; **sub . . . clausa** shut under, or shut at the bottom of: for this use of **sub,** compare Virgil, *Aeneid* 4, 332 'cūram sub corde premēbat').

Suggestions on how to treat the poems in class

We have said that the short poems of Catullus are more likely to evoke a personal response from young readers than anything else in ancient literature. How are they to be treated in class? One axiom is that the teacher should not attempt to tell his class *ex cathedra* how they should respond to a poem; rather the class must feel its way to response through a process of question and answer.

The first task is to establish the sense of the poem. Here it is best for the teacher to begin by reading the whole poem, or if it is too long, the first half of the poem, aloud in Latin. To do this successfully he must be thoroughly at home with the metres of Catullus. We give an appendix on all the metres which occur in our selection on pp.101–3 of the Teacher's Book; in this reading the rhythms of the poem must come across clearly. Then the sense must be established either by translation or by asking comprehension questions. If translation is preferred, it often pays to allow any member of the class to chip in, rather than putting on one individual. If comprehension questions are used, the teacher must be at pains to ensure that the Latin is fully understood at sense level.

When this has been done, the poem should be read aloud in Latin again before discussion begins.

Teachers will not at this level be attempting to give or to elicit a critical exegesis; the terms of criticism need not be used. Our aim is to help pupils respond by ascertaining that they understand what the poet is saying and what emotions he intends to convey. For instance, in the first poem of our selection (31, **Paene īnsulārum . . .**), the key question is 'Why is Catullus so happy?' or 'What does Catullus feel on returning home?' Other questions might be 'What is he glad to be rid of?' 'What does he particularly like about coming home?' Always ask pupils for reference to the Latin to support the views they offer.

Simple questions on structure are often helpful, e.g. 'How does the poem begin?' (by greeting Sirmio, ll.1–3); 'How does it end?' (by greeting Sirmio again, – **salvē** – and by telling it to laugh for joy). Questions on imagery are often appropriate, e.g. **solūtīs cūrīs**; the cares are like a heavy pack which the mind unties.

At the end of such discussion one might hope that the class would not only have formed a personal response but would have begun to see how the poet achieves his purpose. Catullus's poems are not the spontaneous outbursts of emotion which they often appear but expressions of emotion under conscious intellectual control.

Practical criticism of the sort suggested above is time-consuming and we do not suppose that you will be able to treat every poem in this way. But if you are successful with some of the poems, this will influence the whole attitude of your pupils to their reading of ancient (and modern) poetry.

NB In the following notes we refer to the poems by their proper numbers.

Chapter 9

Catullus and his friends

l.1: we give Jerome's date of 87 BC for Catullus's birth, which may not be correct.

Verona was no more than a hill fort before the invasion of the Cimbri (102 BC); after their defeat Transpadane Gaul was settled by Roman colonists, amongst whom may have been Catullus's father or grandfather. Verona rapidly became prosperous and Catullus's father must have been one of its leading citizens, seeing that Julius Caesar was a friend of the family and remained so even after Catullus had attacked him in several obscene poems.

Poem 31

Metre: limping iambics (see Teacher's Book, p.102).
For Catullus's tour of duty on the staff of the governor of Bithynia, see Chapter 11.

The poem is called by Merill (1893) 'a most unartificial and joyous outpouring of feeling'. Certainly no reader could fail to respond to the joy and relief expressed in the central part of the poem in the most natural language. But the poem is not entirely 'unartificial'. The first three lines are sophisticated in expression, especially in the conceit of **uterque Neptūnus**, which probably means 'Neptune in both his capacities' (compare Ovid, *Met.* 1, 338 'lītora sub utrōque iacentia Phoebō', i.e. the sun at its rising and the sun at its setting); and at the end we have another piece of artificiality in **ō Lȳdiae lacūs undae**; Wiseman (p.110) suggests that as the Roman colonists were comparatively recent, they equipped themselves with a legendary past to express their pride in their new country. Certainly the language of the opening and ending is different from that of the central section, more elevated and poetic, perhaps; twice in the last three lines we find **ō**, which is used in Latin only in passages of strong emotion. The only other artificiality in the poem occurs in l.5: **Thȳniam atque Bīthȳnōs/ campōs**, which appears to be a piece of word play; Herodotus (1,28) refers to the Thynoi and Bithynoi as Thracian tribes living in Lydia but the area was known only as Bithynia in Catullus's time.

Poem 50

Metre: hendecasyllables.
l.32 **lūsimus** = **1** we enjoyed ourselves; **2** we wrote amusing verse. (cf. Ovid (*Trīst.* 4, 10, 1) who describes himself as 'tenerōrum lūsor amōrum').
l.33 **dēlicātōs**: the basic meaning is 'delightful', 'charming'. Overtones are: **1** 'sophisticated', 'precious', 'smart', almost 'decadent'; **2** something like 'risqué' (Quinn).
l.34 **versiculōs** little verses = short poems, epigrams.
l.48 **cave**: scanned **căvĕ.** Catullus is warning Calvus not to reject his request for a second meeting; the language of love is continued.
l.51 **cavētō**: the emphatic or so-called future imperative, used in legal and solemn contexts; the poem ends on a note of mock solemnity.

★ p.94: this fresco portrait of a young girl holding tablets and a pen is in the Archaeological Museum in Naples. Discovered at Pompeii, it is popularly known as a portrait of Sappho.

A letter in verse (a favourite form of Catullus) in four sentences; the first gives the reader the information he needs to know; the second (ll.37–43) describes the feelings the evening excited in Catullus; the third explains the reason for the poem; the last (ll.48–51) requests another meeting.

The poem appears completely spontaneous but is carefully constructed (the first six lines would be unnecessary in a real letter).

At l.37 Catullus starts using the language of love (for this language used in a very different context, compare Virgil, *Aeneid* 4 *passim*, especially ll.68ff.). Your pupils may come to the conclusion that Catullus is gay, though the more discerning will see that he is giving a parody of love poetry. The irony of the passage is established in l.39 – unhappy lovers were expected to go off their food but Catullus has only just got back from a dinner party; and in the following lines the symptoms of unrequited love are piled up rather too thickly, ending with the mock solemnity of the final warning.

The poem illustrates in simple form the use of both irony, which is common in Latin poetry, and ambivalence – Catullus's use of the language of love is ironical but this does not mean that he is not serious in his desire to see Calvus again.

Poem 53

Metre: hendecasyllables.
An epigram of which the point comes in the last line.
P. Vatinius was the tribune who in the year of Caesar's consulship put through the **concilium plēbis** the legislation which Caesar wanted. As a henchman of Caesar he became extremely unpopular and was attacked by Cicero in 56 BC. But Cicero was compelled by the triumvirs to defend him on a bribery charge in 54; this was probably the occasion of the poem, when Calvus was prosecuting Vatinius for the second time. The anecdote **num sī iste disertus est . . .** is told by the Elder Seneca (*Cont.* 7, 4, 6).

The point lies in the last line, in which unfortunately the key word, **salapūtium**, is obscure; Bickell's suggestion that it means **mentula salāx**, may be on the right lines.

Poem 14

Metre: hendecasyllables.
Another letter in verse, in which Catullus pretends furious indignation at the gift Calvus has sent him on the eve of the Saturnalia, an anthology of bad poetry. The poem is also an oblique attack on the sort of poetry which Catullus and his circle despised – epic in the old Roman tradition (compare his attack on the *Annals* of Volusius in Poem 36 – **annālēs Volusī, cacāta carta** etc.)

The structure of the poem is rather more sophisticated than that of Poem 50; he plunges straight into his indignation at Calvus's gift (68–70); then asks why his friend should have done such a thing (71–8); a further outburst of indignation (79–83) is followed by a threat to pay him back in kind (84–87), and the poem ends by sending the **pessimī poētae** packing.
l.70 **Vatīniānō**: on Vatinius, see notes on previous poem.
l.73 **clientī**: **cliēns** means a client in relation to a **patrōnus**; it does not have its modern sense.
l.75 **repertum** discovered: suggests that it took some trouble to think up such a gift.
l.76 **litterātor**: an elementary schoolmaster – a rather low form of life at Rome.

★ p.96: a **litterātor** holds one scroll while his pupil studies a second. This marble relief from Veii is from the second or third century AD. (Harrow School)

l.81 **continuō**: best taken as an adverb = straight away, on the spot; but it could agree with **diē** = all day long (a lingering death from boredom).
l.87 **hīs suppliciīs** with this torture, i.e. boredom.
l.89 **unde malum pedem attulistis** = **1** from where you started your ill-omened journey; **2** from where you limped (implying bad metrical feet).

You might ask your class how serious Catullus is in his indignation against Calvus. How does he intend to take his revenge? Did Calvus send the book thinking it was good poetry or as a joke? (Opinions divide on this question but Catullus certainly writes as if he thought it was a joke, cf. **salse** l.83, and we know that Calvus shared Catullus's poetic tastes.)

Poem 93

Metre: elegiac couplets.
Catullus attacks Caesar in three other poems (29, 54, 57), which are all obscenely abusive; according to Suetonius (*Dīv. Iūl.* 73) they were

later reconciled. Caesar invited Catullus to dinner and continued to enjoy the hospitality of Catullus's father.
albus an āter homō: proverbial, compare Cicero *Phil.*2, 41 'quī albus āterve fuerit ignōrās.'.

Poem 49

Metre: hendecasyllables.
An effusive letter of thanks to Cicero for some service he had done him. Quinn suggests that it may be ironical; the language of the first four lines may be felt to go over the top and, as Quinn says, Catullus certainly did not believe that he was the worst of poets. Catullus returned from Bithynia at about the time of Cicero's defence of Caelius Rufus (April 56 BC); on this case see postscript to the following chapter. If Caelius stole Clodia from Catullus and if the occasion of the poem was Cicero's successful defence of Caelius in a prosecution initiated by Clodia, the poem may indeed be ironical; and as Cicero destroyed Clodia's character in his speech, there may also be a note of bitter exultation underlying the extravagant language.

Poem 9

Metre: hendecasyllables.
Veranius had been in Spain with Fabullus, to whom the next poem in our selection is written. They had sent Catullus a present of Spanish napkins (see Poem 12). The poem reads like an unpremeditated outburst of joy, but is carefully constructed, so that the reader knows enough about the situation to share Catullus's feelings. 'A good example of lyric form under conscious intellectual control' (Quinn).

Poem 13 (Exercise 9.3)

Metre: hendecasyllables.

⋆ p.101: this red-figured **kylix** (drinking cup) made in Athens *c.* 490–80 BC shows a drinking party. A youth and a bearded man recline on two couches. With them are women entertainers. A youth enters from the left holding a musical instrument. (British Museum)

A letter addressed to Fabullus inviting him to dinner on his return from Spain. The poem looks like a parody of the invitation poem; Philodemus of Gadara, a contemporary of Catullus, wrote an elegiac epigram inviting his patron to dinner, saying that the food and wine won't be up to much but that the conversation will be good (*Anth. Pal.* 11, 44). Later, Horace writes to invite Maecenas to dinner but warns him that he cannot expect the best wines (*Odes* 1, 20, see Part II, Chapter 17). Catullus goes one better (or worse); instead of a poor dinner, the guest will get no dinner at all, unless he brings it with him.
l.9 **merōs amōrēs**: Quinn translates 'something you will absolutely fall in love with' (i.e. the scent) and quotes an echo from Martial (14, 206), where **merōs amōrēs** is used to denote an object.
ll.11–12 **quod . . . Cupīdinēsque**: Quinn suggests that the party will be a foursome and that the scent will not be that which any Roman host would provide for his guests, but the aura which emanated from his mistress. Compare Virgil, *Aeneid* 1, 402–4, where Venus is described as she leaves Aeneas: 'ambrosiaeque comae dīvīnum vertice odōrem/spīrāvēre' (and her ambrosial locks breathed out a heavenly perfume from her head). This interpretation certainly gives real wit and point to the line.

⋆ p.102: a Roman scent-bottle of gold-banded glass, first century AD. (British Museum)

Chapter 10

Catullus in love

l.2: the identification of Catullus's Lesbia with Clodia, the sister of Publius Clodius, is made by Apuleius (fl. 150 AD) as if it were uncontroversial and it has been accepted by most scholars since. Clodius had two sisters and we assume that Lesbia is the notorious Clodia, wife of Metellus Celer; but some modern scholars, e.g. Wiseman, consider that she was more probably his other sister. It is common practice to treat the Lesbia poems as if they were autobiographical but this is by no means certain. The arrangement of the poems in a biographical sequence seems to us a plausible reconstruction and it would be a very austere teacher who would reject such a gambit in introducing Catullus's poetry to young pupils.
ll.3–4 **virī**: her husband was Q. Caecilius Metellus Celer, consul 60 BC, died 59 BC, poisoned by Clodia, according to Cicero (*Pro Caelio* 59).
l.5: this apostrophe, a common figure in verse,

may strike the reader as a little odd in prose.

Poem 2

Metre: hendecasyllables.
Despite its apparent simplicity, this is a carefully constructed and sophisticated poem. The sparrow was the bird of Aphrodite (Venus); in one of Sappho's most famous poems (Sappho 1), Sappho calls on Aphrodite to come down from heaven in her chariot drawn by sparrows and fulfil her heart's desire.

The poem is written in one long sentence; **passer** is vocative and the following seven lines all qualify this vocative. Catullus invokes the sparrow, as one would invoke a god, and then describes it in relation to **mea puella** in a triad of relative clauses, each increasing in length.

In ll.13–16 both text and meaning are in dispute. With this text, **cum** = when: 'when it pleases the radiant object of my desire to play some sweet game, as a comfort to her pain'; **sōlāciolum suī dolōris** is a second object of **iocārī**; **et** is an explanatory **et**. Wiseman, pp.137ff. takes **nitentī** as agreeing with **meae puellae** (supplied) and **dēsīderiō meō** as ablative – 'shining (i.e. her eyes bright) with longing for me'. Her pain is caused by her longing for Catullus; or is it? **crēdō**, as Wiseman says, is the crucial word. Neither lover has yet declared his/her love.

With the main clause beginning at l.17, the focus of attention shifts from the sparrow and its mistress to the poet. He cannot play with the sparrow (the bird of Venus) and cannot relieve the sad cares (**cūrae** = the cares of love) of his heart.

★ p.104: this vase-painting of two girls playing with a pet bird is in the National Archaeological Museum, Athens.

Poem 5

Metre: hendecasyllables.
vīvāmus . . . amēmus: 'we' conveys togetherness – 'we two against the world' (Wiseman), unmoved by the disapproval which their illicit love provokes in their puritanical elders.

The impact of this poem seems to come partly from the positioning of the words: **vīvāmus . . . amēmus** – first and last words in an endstopped line. **sōlēs . . . nōbīs** emphatically placed at the beginning of lines which each form a complete sentence. **lūx**, a striking monosyllabic line ending, juxtaposed to **nox** at the start of the next line.

The directness and economy of Catullus makes both translations look pretty poor stuff. The comparison of translations is a stimulating but time-consuming exercise, which leads pupils unawares towards an understanding of criticism. Allowance must be made for the date of the translations, e.g. in Campion (1601) 'sager' and 'weigh' may be quite natural expressions (?) but 'Heaven's great lamps do dive . . .' seems a wordy and inadequate rendering of Catullus and misses the point of **occidere**, which means at once 'to set' and 'to die'.

★ p.104, lower illustration: this Hellenistic sculpture shows Cupid embracing Psyche. It is in the Capitoline Museum in Rome and is known as 'The Capitoline Kiss'.

Poem 109

Metre: elegiac couplets
This is the second poem in which Catullus refers to himself and his mistress as **nōs**, but significantly in line 38 the person changes to 'she'. Your pupils should notice this.
aeternum hoc sānctae foedus amīcitiae: 'this eternal compact of hallowed friendship' (Wiseman). Catullus saw their love in a very different light from Clodia, who, he knows in his heart, will not prove capable of keeping such a compact, whatever promises she may make.

Poem 70

Metre: elegiac couplets.
This may have been written before or after Metellus's death; if before, divorce was easy enough in Rome. The form of the poem owes much to an epigram by Callimachus (*Ep.*25) but whereas Callimachus is light-hearted and ironical about the promises of lovers, Catullus is deadly serious.

Poem 85

Metre: elegiac couplets.
The answer to the question may be that the poem makes its impact (1) by the extreme simplicity and directness of its statement; (2) by the shift from the active (**faciam**) to the passive (**excrucior**) (Catullus is not doing anything but having something done to him); and by the weight of the final word, the only word which is

not part of everyday language. Ovid (*Amōrēs* 3,11b) treats the same theme ironically in a most amusing poem which begins:

luctantur pectusque leve in contrāria tendunt
hāc amor, hāc odium; sed, puto, vincit amor.

Poem 8

Metre: limping iambics.
The poem starts as a soliloquy in which Catullus exhorts himself to stop being a fool. He looks back on the sunlit past of their love with infinite sadness, but then pulls himself up and tells himself to harden his heart.

The second section is addressed to his girl – when Catullus leaves her, she will be sorry, but the series of questions becomes increasingly emotional as they recall his own past love.

In the last line he again pulls himself up – he must forget these thoughts and be strong.

Throughout the poem runs the antithesis between Catullus's intellectual decision to leave his mistress and his emotional inability to do so.

Poem 76

Metre: elegiac couplets.
ll.75–82: the poem appears to start calmly as Catullus reassures himself that he will have consolation, since he has been **pius**, has kept the **foedus** he made with his mistress, and done and said all that any man could. But Catullus is developing an old commonplace (see, e.g. Cicero, *Dē Senectūte* 9 'aptissima omnīnō sunt arma senectūtis artēs exercitātiōnēsque virtūtum, . . . cōnscientia bene āctae vītae multōrumque benefactōrum recordātiō iucundissima est'. The most appropriate weapons of old age are the study and practice of the virtues . . . because the consciousness of a life well led and the memory of many good deeds is most pleasant).

The conditional – **sī qua . . . voluptās est hominī** – points the irony and bitterness which underlie his statement of a well-worn theme.
ll.83–90: the emotional intensity rises, as he argues with himself in his efforts to shake off this love. We know he cannot do it.
ll.91–100: at l.91 he bursts into a desperate prayer to the gods to tear from him this destructive plague; now he only wishes to be free.

This powerful poem is again written in language of extreme simplicity, mostly in the rhythms and diction of everyday speech.

Poem 3 (Exercise 10.4)

Metre: hendecasyllables.
The poem takes the traditional form of a dirge: the call to mourn the death of the departed; a **laudātiō** of the dead (developed in a triad, each member increasing in length – one line, two lines, three lines); the journey to the bourn from which no traveller returns. This is followed by an outburst of indignation against the destructive powers of Death; and the poem ends with a shift of focus from the sparrow to the poet's mistress.

In using this form Catullus sounds a note of irony from the very first line – the solemn dirge is hardly appropriate mourning for the death of the most diminutive of birds. And those called to mourn are not e.g. 'all good men', but 'all powers of love' and 'all men devoted to love' (**venustus**, a favourite adjective of Catullus, had come to mean 'charming', 'refined' etc. but in this context its connection with Venus must surely be its leading connotation). The **laudātiō** of the sparrow dwells only on its relations with **mea puella**. The solemnity of the description of the sparrow's lonely journey down to the darkness of death is belied by the sound effects; it begins with four heavy monosyllables (suggestive of the sparrow hopping) and an astonishing jingle: **it per iter**; and this is followed by a line of the most impressive solemnity. The sparrow is acting on a stage which is too grand for it, just as the curse on Orcus is out of all proportion to the situation.

The sudden switch to **mea puella** in the last two lines makes explicit what was suggested by the first line – that this is a love poem rather than a dirge.

Clodia and Caelius

There may not be time to treat this fully but it might be worth translating Poem 77 and the passage from the *Prō Caeliō*.

The identification of Rufus with M. Caelius Rufus is not universally accepted; there are other poems to a Rufus (69 and 71), who had smelly armpits and does not sound a bit like the fashionable Caelius. But as a Caelius certainly did succeed Catullus as one of Clodia's lovers, it

seems extremely plausible that he is the subject of this poem.
ll.9–10: **venēnum** and **pestis** both appear to be vocative. Rufus is the 'cruel poison of my life' and 'the destruction (destroyer) of my friendship (or 'love')'.
l.13 **lītem**: the prosecution was brought by L. Sempronius Atratinus, whose father Caelius had prosecuted the year before, in April 56 BC. The charge was riot (**dē vī**). But whatever the technical charges were, it is clear that Clodia was the driving force behind the prosecution and that the case was brought from personal motives.

The quotation is from *Prō Caeliō* 49–50.

★p.115: this drawing is closely based on a wall painting from the **triclīnium** of a house in Pompeii. It is the second of three related panels depicting dinner parties, possibly successive stages of one party. The first shows a respectable scene, with the guests reclining fully clothed on their couches; the third includes musicians and nude dancing girls. The second scene takes place in a garden under an awning. In the foreground is a table set with food and wine; on the right a boy approaches bringing more wine in **askoi**. On the left-hand couch are a couple stripped from the wait upwards; the woman raises a drinking horn to her mouth, the man embraces her with his right arm and holds a plate in his left hand. On the central couch reclines the host, who says: FACITIS·VOBIS·SVAVITER·EGO·CANTO (Enjoy yourselves; I'm singing). On the right reclines another couple; the man says: EST·ITA·VALEAS (All right; good luck to you).

FACITIS: this might be intended as an indicative (you are enoying yourselves) or, more likely, it is an error on the part of the artist, who intended to write FACIATIS (jussive subjunctive). Your pupils might be comforted to know that grammatical mistakes are not uncommon in inscriptions. The captions illustrate typical Roman writing; all in capitals, the words here separated by dots, but no punctuation; no letter U — SVAVITER.

Chapter 11

Catullus – the sequel

Poem 11

Metre: Sapphics.
This, which is generally considered the last of the Lesbia poems, is written in Sapphics, the metre Catullus used for Poem 51, which is believed by many to be the first.

The poem is cunningly constructed with two violent changes of direction and shifts in tone.

The romantic travelogue (ll.3–15) leads the reader, or listener, up the garden path; he can have no idea of what is to follow **temptāre simul parātī**; the most dangerous mission of all turns out to be:

pauca nūntiāte meae puellae

and not until the next line do we learn why this mission is so dangerous.

The fifth stanza gives the message in terms of bitter contempt, in which sound effects are expressive of the sense: alliteration – **vīvat valeatque, tenet trecentōs** – and in the following line three elisions of syllables ending in m, one over the end of the line (a rare license):
nullum amāns vērē, sed identidem omnium . . .
The interpretation of such an effect is apt to be subjective but to our ear it is extremely ugly, a sort of slimy slur.

In the last stanza there comes another complete shift in tone, as he looks back on his love as a thing of beauty, ruthlessly and carelessly cut down by her fault. Here there is a sound effect – the elision of **prātī** over the end of the line – of which we can speak more confidently; the cutting off of this syllable expresses the cutting off of the flower.

★p. 117: this Scythian bronze is in the British Museum.

Poem 46

Metre: hendecasyllables.
Structure: ll.30–5 Spring is here; it's time to be off.
ll.36–40 I long to start; friends, goodbye.
The linking line, in our view, is **ad clārās . . . volēmus**.

Our reason for preferring this interpretation of the structure to that of e.g. Quinn lies in the sound echoes:
l.30 **iam vēr ēgelidōs refert tepōrēs**

l.35 **iam mēns praetrepidāns avet vagārī**
(apart from the sound echoes in the first two words of each line, the lines are rhythmically identical)
l.31 **iam caelī** . . .
l.36 **iam laetī** . . .

Catullus's state of mind echoes the state of the season – new growth, new beginnings after storms. A poem of joy and relief, just tinged with sadness at the end, as he will miss his friends.

★ p. 118: St Paul's Christian teaching sparked off a serious demonstration here (*Acts*, 19,22ff.).

★ p. 119: a wall-painting from Pompeii.

Poem 4

Metre: pure iambic trimeters.
After the opening line the whole poem except for the last three lines is put into the mouth of the yacht, a boastful and garrulous old boat, which traces its life backwards to its first beginnings as a tree on the hills of Bithynia.

The poem is a metrical *tour de force*; to write pure iambics (with no use of spondees) in a language as heavy as Latin requires extraordinary skill and yet it reads as if it were the easiest thing in the world.

Poem 10

Metre: hendecasyllables.
This is the best example of Catullus's mastery of the use of colloquial diction in verse; throughout the poem we feel the language and rhythms of rather racy everyday speech. The only poet to manage this skill so well was Horace in some of his *Satires* and *Epistles* (e.g. *Satire* 1, 5 and 9; see Part II, Chapters 15 and 16).

★ p. 122: this sculpture of Jupiter Serapis dates from the second century AD. It was found in Rome in 1775 and is now in the British Museum.

The dative of the person concerned

Under this heading we have brought together several usages to which grammarians give different names (possessive dative, dative of advantage etc.). These usages can only be understood from experience and it seemed to us better not to introduce too many grammatical names. We continue to gloss hard instances as they occur. The hardest is probably of the type

ēripite hanc pestem mihi

where one might expect an ablative.

Poem 101 (Exercise 11.3)

Metre: elegiac couplets.
We learn from Poem 68 that Catullus's brother died at Troy:

Trōia (nefās!) commūne sepulcrum Asiae Eurōpaeque,
Trōia virum et virtūtum omnium acerba cinis,
quaene etiam nostrō lētum miserābile frātrī attulit. ei miserō frāter adēmpte mihi,
ei miserō frātrī iūcundum lūmen adēmptum,
tēcum ūnā tōta est nostra sepulta domus,
omnia tēcum ūnā periērunt gaudia nostra,
quae tuus in vītā dulcis alēbat amor.
(ll.89–96)

The poem's strength is partly due to the directness of the thought and the dignified simplicity of the language. When you read the last of the Ovid poems, the Elegy on Tibullus (p.224), it would be instructive to compare the two poems. Ovid's elegy is another powerful expression of grief, more complex in thought and more decorated in expression.

Chapters 12–14 Virgil

Chapter 12 consists of excerpts from Book 1 of the *Aeneid*, Chapters 13 and 14 of excerpts from Book 4. A good modern edition of *Aeneid* 1–6 is that edited by R. D. Williams, Macmillan. R. G. Austin's editions of Books 1, Oxford, and 4, Oxford, are highly recommended. See also T. E. Page's edition of *Aeneid* 1–6, Macmillan. A valuable and readable book on the poet's work as a whole is Jasper Griffin's *Virgil*, Oxford.

The first three words of Virgil's *Aeneid* – **arma virumque canō** – not only introduce the story of the warrior Aeneas, but are also a declaration of poetic intent. **arma** evokes Homer's *Iliad*, the first poem of European literature. Its subject was the fighting round the city of Troy, and in it the Trojan prince Aeneas plays a minor but significant role. **virum** calls Homer's *Odyssey* to mind. The first word of that poem is *andra* (man) and it tells of the hero Odysseus's adventures as he travels from the

Trojan War back to his island of Ithaca and re-establishes himself there as king.

By making his readers think of Homer right at the outset, Virgil shows astonishing ambition in setting his work on a level with his great predecessor in epic. He puts his hero in the same world as Achilles, Hector, Priam and the other great figures of the Trojan War, and he adds a Homeric dimension to the travels of Aeneas (the first half of the *Aeneid*, Books 1–6) by modelling them on the *Odyssey*, and to the dreadful war Aeneas undergoes in Italy (the second half, Books 7–12) by basing it on the *Iliad*.

But the differences between Homer and the *Aeneid* are as important as the similarities. Odysseus, for example, is travelling back to his homeland and his wife. Aeneas's home of Troy lies in ruins and he must journey towards a mysterious future and a city and empire which he will never see. Odysseus loses all of his companions and arrives at Ithaca alone. Aeneas is a leader of a new kind with a social responsibility; **pietās** is the key to his character; he only briefly forgets his duty to his family, his gods and his men. The endlessly inventive Odysseus revels in the challenging dangers which confront him in a hostile world. Aeneas's destiny involves him in labours which he undergoes doggedly. Odysseus loves his wife and manages to part with his mistresses on friendly terms. Aeneas, on the other hand, loses his wife and, when he leaves Dido, she cannot forgive him and curses him with disastrous repercussions for their descendants.

The correspondences and dissimilarities between the epics of Homer and Virgil are a fertile field for debate, and they are much more complicated than we have suggested, for Virgil makes copious use of Homer's language as well as his themes.

There are further debts. We give here the story of Dido and Aeneas, which Virgil did not invent but treated with unprecedented depth and pathos. The most important sources for this part of the poem are **1** Apollonius of Rhodes who wrote with sensitivity and intensity of the love of Medea for Jason in his *Argonautica (The Voyage of the Argo*, third century BC), **2** Catullus, who had responded with a poignant sense of personal involvement to the situation of Ariadne abandoned by Theseus (Poem 64), and **3** Euripides with his portrayals of wild emotion and his mastery of feminine psychology. For the story of Dido and Aeneas, recent history provided a parallel, since, as Augustan propaganda had it, the fatal Oriental queen Cleopatra enslaved and destroyed the great Roman leader Mark Antony.

Virgil wrote the *Aeneid* between 29 and 19 BC. He died before he could finally revise it.

For the Virgilian hexameter, see Appendix, pp.102 and 103.

Chapter 12

Aeneas arrives at Carthage

Aeneid 1, 12–22, 29–33, 338–66, 418–38, 494–508

Introductory passage: you may wish to tell your pupils the story of the Trojan horse.
l.1 **antīqua**: the city of Carthage is still being built when Aeneas reaches North Africa. Thus it is **antīqua** not from Aeneas's point of view but from that of Virgil. At many points in the *Aeneid*, Virgil makes such use of 'double time', viewing events both as they occur in his narrative and as they appear in the context of Augustan Rome. Between the times of Aeneas and Virgil the Punic Wars had been fought between Carthage and Rome, ending in the total destruction of Carthage in 146 BC. See Chapters 15–17.
ll.2–3 **Italiam . . . Tiberīnaque ōstia**: as we suggest in our gloss, this means 'Italy where the Tiber runs into the sea' rather than 'Italy *and* the mouth of the Tiber'. This kind of phrase, where the second unit explains the first, is common in Virgil and other Latin poets. **Tiberīna ōstia**: 'the mouth of the Tiber' but also Ostia, the harbour located there.
l.5: Juno's love for Carthage is due to the fact that she was worshipped there as the goddess Tanit.
l.11 **Parcās**: the three Fates were Clotho (the Spinner – hence **volvere**), Atropos and Lachesis. What is fated in the *Aeneid* will happen but it can be delayed or modified. Vulcan tells Venus that he could have made the Trojan War last for another ten years; neither the Fates nor Jupiter would have stopped him (8, 398–9).
l.16: a famous line. The labour and suffering involved in establishing the Roman race are repeatedly stressed in the poem. **Rōmānam condere gentem**: this is Aeneas's destiny.

★ p. 130 (right): Juno holds a sceptre in her right hand and a phial in her left. The statue is known as the Hera Barberini and was found on the Viminal Hill in the seventeenth century.

★ p. 130 (left): the Vatican Library has in its collection the two earliest illustrated manuscripts of Virgil, both dating from the close of the Classical period. This illustration of the storm comes from the later of them, the *Cōdex Rōmānus*, fifth or sixth century AD. Close to despair, Aeneas stretches forth his arms as malign figures shower rain and lightning from above. Beady-eyed sea monsters gleefully await their prey.

l.27 **furor**: a key concept in the poem: it is opposed by **pietās**, the quality possessed above all by Aeneas.
l.28–30 **impius**: see last note. The impiety of Pygmalion is stressed both by the scene of the murder (**ante ārās**) and by his violation of the family bond in killing his sister's husband.
l.32 **inhumātī**: the impious Pygmalion has not even buried his victim. Without burial, the spirits of the dead could not find rest.
l.34 **crūdēlīs**: this is the first instance we have met of a 3rd declension accusative plural ending in **-īs** instead of **-ēs**. This occurs very frequently. In this chapter every occurrence is glossed, but subsequently we expect the form to be recognized.
l.39 **fugam . . . sociōsque parābat**: a good example of syllepsis. The verb has its literal meaning with one of its objects and a non-literal meaning with the other: cf. 'He lost his hat and his temper.' In English syllepsis is often used to humorous effect, but rarely so in Latin.
l.43 **dux fēmina factī**: throughout this passage (ll.19–45), Virgil has shown to what extent Dido is an 'alter Aenēās' (a second Aeneas). They have both lost their partners in marriage; both of them, warned by a ghost in a dream, have fled from a city where there is no future for them, showing outstanding courage and leadership. Aeneas wishes to found a new city; Dido is already founding one. With so much in common, it is not surprising that they are attracted to each other.
ll.49–50 **mīrātur . . . mīrātur**: Aeneas is suitably impressed by the construction of the city. City-building is an important theme of the poem; **māgālia** is a Carthaginian word. T. E. Page suggests that 'Virgil is probably thinking of the view of Rome from the Esquiline, from the palace from which Horace tells us that Maecenas loved 'mīrārī beātae/fūmum et opēs strepitumque Rōmae' (*Odes* 3, 29, 11–12).

★ p. 132 (upper): this illustration from the earlier (fourth century) Vatican manuscript, the *Cōdex Vāticānus*, shows Aeneas gazing in wonder at the building of Carthage. The wheel is part of a crane (**māchina**, p. 138, l. 30).
l.54 **senātum**: Carthage had a senate from about 400 BC, but the civilized order shown in Dido's projected city indicates that Virgil is creating an ideal picture rather than trying to describe the beginnings of the real Carthage. It is the city of Aeneas's dreams.
ll.58–64: Virgil elsewhere uses bees to convey a co-operative and happy community. See *Georgics* 4 *passim* and *Aeneid* 6, 707–9. Brooks Otis remarks that 'the simile suggests all the sweetness of security and happy employment' and W. R. Johnson finds in it an expression of Virgil's 'own best dream, the unity of the City'.
l.65: 'The want of a city is the keynote of the Aeneid.' – J. Conington.

★ p. 132 (lower): this bee is on a silver coin minted in Ephesus between 387 and 295 BC.

Exercise 12.2

ll.5–9: the simile is based on Homer, *Odyssey* 6, 102ff. It is altogether joyous, but note that Diana **pharetram/fert umerō**; i.e. she is a huntress. Dido, on the other hand, is to prove the victim of the hunt.
l.12 **foribus dīvae**: the doors of the shrine (**cella**) at the back of the temple's main hall which has a vaulted roof (**testūdine**).

★ p. 135: the Basilica of Constantine (fourth century AD) was modelled on the great halls of Roman bath houses: its three arches still dominate the forum. 24½ metres high and originally faced with marble and stucco, it is a sensational example of concrete construction. Michelangelo is said to have studied the building when dealing with the problems of the dome of St Peter's.

Question 4: make sure that the 3rd declension accusative plural ending **-īs** has been mastered.

★ p. 136: this statue, a copy of a work by Praxiteles, is in the Capitoline Museum, Rome.

★ p. 137: from the *Cōdex Rōmānus*, the later Vatican manuscript. Dido reclines on a semi-circular couch between Aeneas (on the left) and Achates. The Trojans wear Phrygian caps and purple robes. The haloes show a curious melding of Christian and pagan art.

Chapter 13

Dido and Aeneas

Aeneid 4, 1–5, 65–89, 129–36, 138–72, 259–86, 305–30

ll.1–2: Cupid's traditional weapons are arrows (causing the wound of love) and fire (representing the flames of passion). At the end of Book 4, Dido stabs herself and then (5, 2–3) is consumed by the flames of a pyre. The metaphors become realities; **iamdūdum**: Dido has in fact fallen in love very quickly, through divine intervention; the use of this word conveys her impression that her love for Aeneas has already lasted a long time; **cūrā** suggests the anxious pain of love. Throughout the book, Virgil presents Dido's love as a torment, as something that consumes her. Contrast the usual modern view of love (propagated above all by Shakespeare) as life-enhancing.
l.5 **nec . . . dat cūra**: be careful not to allow **nec** to be translated as 'nor', which will lead to translator's language and not real English: 'and her suffering withholds . . .'.
ll.7–9: the prophets and the rituals Dido performs are useless. Her suffering lies deep within her. The ideas of fire (**flamma**) and the wound (**vulnus**) should be stressed; **tacitum vulnus**: cf. *Hamlet* 1,2: 'But break my heart for I must hold my tongue.'
ll.10–14: analysis of this celebrated simile must not do less than justice to its pathos and its evocation of pain. Note its location in the cold and wild uplands of Crete. Aeneas is the shepherd who is out hunting. (Is there a contradiction between the roles of shepherd and hunter?) He shoots an arrow at a doe he is pursuing and is unaware that it has hit her. (Is Aeneas in any way guilty?) The doe is Dido, caught off her guard, shot, and then roaming restlessly over the mountain slopes in dreadful agony. The **sagitta** is the arrow of love which cannot be removed (**haeret**). Matthew Arnold poignantly evokes this simile in *Sohrab and Rustum* 503–6:

> most like the roar
> Of some pain'd desert-lion, who all day
> Hath trail'd the hunter's javelin in his side,
> And comes at night to die upon the sand.

l.13 **silvās saltūsque**: hendiadys; i.e. the two nouns suggest one idea, 'the wooded defiles'. Cf. 'bread and butter'. Explain this to your pupils.

★p. 138 (upper): this mosaic of a deer is from a Roman villa in Tunisia. (Bardo Museum, Tunis)

★p. 138 (lower): this hunter is painted on an Attic red-figure vase at present in Boston. It is the work of the Pan Painter and dates from *c.* 470–60 BC.

l.22: this evocative line should be scanned and its sound savoured. The last three words form a beautiful cadence with ictus (the stress dictated by scansion) coinciding with speech stress. See Appendix, p.103.
l.25: it is not clear from Virgil's narrative when the real Ascanius is brought back to replace the disguised Cupid.
ll.29–30: all work on the city has stopped. The crane simply standing there idle is an evocative symbol of suspended activity. Dido's private passion has taken her over: it has eclipsed her public *persona* as queen of Carthage. The contrast with the **dux fēmina factī** of Book 1 should be stressed.

★p. 134: this mosaic from Carthage dates from the early fourth century AD and shows a boar hunt.

l.31: the river Oceanus emerged from the Underworld and flowed about the earth which was conceived as flat (like a plate).
l.33: in a Roman hunt, nets were set up at the far side of the hunting area; the quarry was driven into these and killed.
There is no verb in this line. The gear comes (**it**, l.32) out at the gates with the young men. The poet's eye leaps excitedly from item to item. Encourage your pupils to translate one 'period' at a time. This is the secret of reading Virgil.
l.34: the monosyllable at the end of the line dislocates the cadence and conveys an impression of bustling activity which is well in keeping with the animation of this passage as a whole. See Appendix, p.103.
l.35: Dido lingers in her bedroom with the bashfulness of a bride about to set off for her marriage.

l.42 **nec nōn et** and . . . as well. **Iūlus** and **Ascanius** are names used interchangeably of Aeneas's son.
ll.44–50: this simile likening Aeneas to Apollo looks back to the simile comparing Dido with Apollo's sister Diana (p.135, ll.5–9). The summer setting with the jovial noise of an international festival and the splendid appearance of Apollo himself create a heady impression. But note the sinister effect of the weapons sounding on Apollo's shoulder (cf. Homer, *Iliad* 1, 45–7). Aeneas and the arrow of love are to prove fatal to Dido. The picture of Apollo balances delicacy with power.

It was traditional to use similes to characterize. Cf. the simile by which Jason is introduced when meeting Medea in the *Argonautica* of Apollonius:

> She longed to see him, and he soon appeared to her, his tall figure springing into view like the dog-star Sirius coming from the Ocean. The star rises beautiful and clear to behold but brings unspeakable woe to the animals. Thus Jason came to her, beautiful to look on, but his appearance aroused the torments of love. (3, 956–61)

★ p. 140 (upper): this statue of Apollo is from a bronze original, probably of the fourth century BC. The god steps forward to see the effect of the arrow he has just shot. The graceful elegance and strength of his body and his keen gaze have won much admiration. The statue is called the Apollo Belvedere because it is placed in the Belvedere of the Vatican.

l.56: to an English-speaking reader it appears that Virgil is saying that the deer first career over the plains and *then* quit the mountains – which is clearly not his meaning. Virgil's descriptive technique is not sequential but cumulative.
ll.57–60: Aeneas is about to embark on a love affair which is to put his son's dynastic future at risk. It is appropriate that Ascanius should be presented here as the epitome of mettlesome youth. Surely we have the real Ascanius back with us by now.

★ p. 140 (lower): a detail from a sarcophagus in the Museo Nazionale Romano, Rome.

ll.61ff.: the storm is emblematic of the forces of **furor** in the world (cf. 1, 148–56).

l.66 **dux**: the use of the word is devastating. By his union with Dido, Aeneas effectively abandons his leadership of his men. He becomes Dido's consort. He *leads* Dido into the cave; and she persuades herself that he is her bridegroom (**dūcō** can mean 'I marry').
ll.67ff.: 'The elemental powers of nature and supernatural divinities conspire to produce a parody of a wedding, a hallucination by which the unhappy Dido is deceived.' – R. D. Williams. Cf. Milton's fine lines when Eve eats the apple:

> Earth felt the wound, and Nature from her seate
> Sighing through all her Works gave signs of woe,
> That all was lost.
> (*Paradise Lost* 9, 782–4)

l.73: on his side, Aeneas has made no commitment.

★ p. 142: from the later Vatican manuscript which is more cartoon-like than the earlier one. As the royal couple make love in the cave, two fellow-hunters shelter under a tree and a shield respectively. One of the horses looks anxiously upward.

★ p. 143: this relief from a Roman marble sarcophagus of the second century AD shows the Roman marriage ceremony of **dextrārum iūnctiō**. In their left hands, the bride holds a rose and the husband the wedding contract. Behind them stands the bridesmaid. (British Museum)

l.74: Mercury wears winged sandals.

★ p. 144: this Giam Bologna statue of Mercury is Renaissance in period but classical in feeling. He has wings on his hat and his ankles and he holds his **cadūceus**, a staff with two serpents twining round it.

l.75: work has started up again. But it is not Carthage that Aeneas should be building.
ll.76–9: this oriental picture of Aeneas, wearing a cloak made for him by Dido, gives an effeminate impression. He is neglecting his Roman duty. **Tyriō mūrice**: Tyrian dye, which produced a glowing purple colour, was made from an extract from a vein of the whelk (**mūrex** a shellfish). It was very expensive with some 12,000 whelks being needed to produce 1.5 grams of fluid.

l.81 **uxōrius**: Mercury is being highly contemptuous.
l.82 **rēgnī rērumque . . . tuārum**: i.e. Aeneas's Roman and Italian destiny.
l.86: note the jingle of **teris** and **terrīs**: does this convey the idea of lingering?
l.95: after his initial shock – on which Virgil focuses arrestingly –, Aeneas is on fire to leave, and quickly (**fugā**). But note that the land of Carthage is still sweet (**dulcīs**) to him.
ll.97–8 **agat . . . audeat . . . sūmat**: deliberative subjunctives.
ll.99–100: Aeneas's mental *impasse* is strikingly conveyed. Cf. Tennyson's 'this way and that dividing the swift mind'.

Grammar

No new syntax in this chapter or the next. Concentrate on the poetry.

Exercise 13.2

ll.16–22: the translation is by C. Day Lewis.
l.24 **parvulus**: the only example of a diminutive adjective in the *Aeneid*. It creates an effect of extraordinary intimacy with the non-existent child.
ll.1–26: the enormous range of emotions spanned in this speech calls for discussion.
You might ask your pupils if they feel that it is very much a woman's speech and, if so, why?
Question 2: presumably she has not yet formulated her plan to commit suicide. Is the meaning simply that she cannot contemplate life without Aeneas? Thus his departure means death for her.
Question 5: she concludes that he is running away from her: **mēne fugis?** (l.10).

★p. 149: the mosaic floor of the **frigidārium** from Low Ham in Somerset (fourth century AD). You can now review the story so far; it is told anti-clockwise from the bottom right. (a) Aeneas' ships arrive at the coast of Africa; (b) (at the top): Venus and Cupid (disguised as Ascanius) supervise the meeting of Dido and Aeneas; (c) (on the left): they go hunting; Aeneas looks back at Dido while Ascanius rushes ahead; (d) (bottom) they make love. At the centre, Venus is flanked by two Cupids holding lowered and raised torches, symbolising the death of Dido and the continuing life of Aeneas respectively. (Achates, holding a gift for Dido, appears to be on his back at the top right: in fact, he belongs to the group (b).) The Trojans wear Phrygian caps.

Chapter 14

The death of Dido

Aeneid 4, 330–61, 393–415, 607–29, 642–705

l.2: Virgil tells us that it was an effort for Aeneas to stifle his love.
l.3ff.: we quote two extreme views of this, the only speech of Aeneas in Book 4: 'Not all Virgil's art can make the figure of Aeneas here appear other than despicable. His conduct had been vile, and Dido's heart-broken appeal brings its vileness into strong relief.' – Page. 'Virgil has taken the utmost care to convey the reasons why Aeneas's reply is cold; it is (1–2) because he knows he must not yield and therefore he smothers his love and his emotions. He endeavours to use logical and persuasive arguments to put his case, honestly believing that Dido will see that he has no option.' – Williams.
l.3 **ego tē**: 'the two persons concerned face each other syntactically, as it were.' – R. G. Austin.
l.7 **prō rē pauca loquar**: usually taken to mean 'I shall speak a few words to meet the case', and this is certainly in keeping with the formal, legalistic tone of the speech. But Aeneas could be saying, 'In proportion to (the greatness of) the matter, I shall speak (only) a few words', i.e. he could be acknowledging the magnitude of what is happening.
ll.7–8: would there in fact ever have been a right time or a right way for Aeneas to break the news to Dido?

★p. 151: Gian Lorenzo Bernini was only 15 when he sculpted this marble group jointly with his father in 1613. It shows the hero as he escapes from Troy, carrying his father Anchises and leading his son Ascanius by the hand. Anchises carries an image of the goddess Vesta, Ascanius holds the sacred fire, and Aeneas wears a lion skin. This was a highly popular subject in the ancient world. Compare I, p. 70. (Galleria Borghese, Rome).

ll.10–14: Aeneas points out chillingly that if he had his way he would rebuild Troy rather than stay with Dido. This is an arresting

statement of the city-building theme. **Priamī tēcta alta manērent**: i.e. he would build Priam's palace again on the ruins of the old one.
ll.15–16, 21–3: we have not heard about these oracles or the appearances of Anchises's ghost before now.
ll.28–9: Mercury appeared to him in person, not in a dream. He stresses this fact.
l.31: Virgil would have completed this and the other half-lines in the poem if he had lived to finish the poem. 'Nothing . . .,' writes Page, 'could improve these four words thus left rugged and abrupt.' But how could Page, or anyone else for that matter, know that Virgil could not in fact have improved on them?
l.33: now that Aeneas is acting in accordance with his destiny, he can have his epithet of **pius** restored to him. It has not been used since 1,378 (this is 4,393). His **pietās** is the reason for – indeed the only justification for – his departure. There is a sense in which Virgil is here emptying the word **pius** of any laudatory or condemnatory associations. He is simply telling us the fundamental truth about the man we now see. Ask your pupils how they respond to this word at this stage of the story.
l.34: Virgil again (cf. 1.2) assures us that Aeneas's feelings were profound, but these almost editorial statements do not carry as great a weight as words and deeds. In fact, we see into Dido's heart in this book in a way that is almost completely denied us in the case of Aeneas. (But see 4, 441–9).
l.35: in embracing the orders of the gods, Aeneas shows himself to be a true Stoic.
l.37: 'The use of the singular here gives a sudden picture of one of the many ships as it is launched.' – Williams
ll.38–9: it is a curious accident that the second of these lines about incomplete oars was itself left incomplete at Virgil's death.
ll.41–6: the simile Virgil uses elsewhere in the poem to suggest a happily co-operative community, that of bees in a sunlit and flowering countryside (p.132, ll.58–64; 6, 707–9), is here transformed into a description of efficient ants. Effortfully provident as ants proverbially are, they summon up a picture of communal discipline for Dido and for us to behold. It is in this bleak light that we are asked at this moment to view the destiny of Rome.

★ p. 152: the ant picture is a cornelian gem of the first or second century AD. (British Museum)

l.43: a good line to demonstrate the clash of ictus with speech stress (Teacher's Book, p.103). There is a sense of laborious trudging. Ennius used **it nigrum campīs agmen** of elephants.
l.44 **grandia**: i.e. from the point of view of the ants.
l.47: Virgil addresses one of his characters in the second person. This is a device called *apostrophe* (the poet 'turns away from' his narrative). Here it indicates his profound sympathy for Dido.
l.49 **arce ex summā**: Dido stands on her lonely height. Far below, the Trojans do indeed look like ants.
l.51: in a second apostrophe, Virgil addresses the god of love. How effective does this prove? Does it intensify the emotional effect or weaken it by protesting too much?
l.55 **Sōl**: the sun sees everything: nothing can be hidden from it. Dido invokes it as a witness of Aeneas's treachery.
l.56: she speaks more truly than she knows. For Juno's role in furthering the liaison between Dido and Aeneas, see p.140. From the reader's point of view, though not from Dido's, **interpres** may here come close to meaning 'bawd'.
l.57: Hecate was worshipped at the crossroads.
l.61 **caput**: denotes contempt.
l.63ff.: Virgil modelled these curses on Polyphemus's curse upon Odysseus in *Odyssey* 9, 534ff., another tale of violated hospitality. The Polyphemus story is very different, but there are some instructive comparisons to be made. While Dido's curses were all fulfilled, some do not happen in the *Aeneid*. The educated Roman reader was expected to have picked up such information from his reading elsewhere.

Charles I was confronted by these terrible lines when he opened his Virgil in the 1640s. He was taking the *Sortēs Vergiliānae*. To do this, you open your Virgil suddenly; the passage you touch at random with your finger is the oracular response.
l.70: she calls for undying enmity between Romans and Carthaginians.
l.71: cf. Catullus's poem about his brother's death (p.126, ll.3–4).
l.73 **aliquis . . . ultor**: this refers to Hannibal who came close to destroying Rome in 217 and 216 BC (see p.185). The use of **aliquis** with the second person is startling. Dido cries out across

the centuries with a passionate immediacy to someone she cannot know. The hissing 's's in this line are only one feature of a speech which contains many extraordinary sound effects.
ll.76–7: the idea of opposition is starkly reflected in the word order. The final **-que** goes over the end of the hexameter (i.e. is hypermetric). 'The never-ending hatred of Dido is reflected in the unended rhythm of her final words.' – Williams. And Austin comments, more subjectively, 'Dido seems to leave the two peoples locked for ever in their enmity.'

★p. 154: this bust of Hannibal is in the National Archaeological Museum in Naples.

ll.79–80: **trementēs/interfūsa genās**: literally, blotched as to her quivering cheeks – **trementīs genās** is an accusative of respect.
l.82: a ladder leans against the pyre to enable Dido to climb to its top.
l.88 **exsolvite**: Dido longs for release.
l.89ff.: Heinze writes of Dido's 'echt römischer Heroismus' (truly Roman heroism). Like Horace's Cleopatra (*Odes* 1, 37, 30–4), Dido becomes the Stoic Roman in death. Her self-assertion has the simple dignity of inscriptions on tombstones.
l.90 **magna . . . imāgō**: ghosts were thought to be larger than life – or perhaps Dido means that in her death she will achieve greatness.
l.91 **urbem praeclāram statuī**: the city-building theme. In this line and the next, Dido reasserts her public self, looking back to the 'alter Aenēās' theme.
l.92: how has she taken revenge upon her brother? She got away with her husband's hidden treasure, though it was partly for this that Pygmalion had killed him; she commandeered some of his ships; and she fled with a considerable number of dissidents (p.132, ll.37–43).
l.94 **carīnae**: synecdoche, the use of part of something to express the whole of it.
l.96: the ancient commentator Servius suggests that as she says **sīc, sīc**, Dido stabs herself twice. Shades of Bottom in *A Midsummer Night's Dream*! Surely Dido simply means, 'Thus, thus it is that I choose to die.'

★p. 156: this beautiful page of a manuscript of the late fifteenth century shows Dido on her pyre as Aeneas sails away. The prelude to the tragedy is shown: as the hunt proceeds, Dido and Aeneas enter the cave. (British Museum).

ll.105–7: the simile of the sack of Carthage is a fine example of Virgil's use of double time. The poet makes his commentary on the ultimate failure of Dido's curse. The conflict between Carthage and Rome was to end with the sack of Carthage in 146 BC. The simile is based on *Iliad* 22, 410. Austin comments that the sacking of a city was 'the most dreadful scene of horror that an ancient writer could imagine'. If we have read continuously through the poem, we still have in our mind the powerful description of the sack of Troy in Book 2.
l.109 **ōra . . . pectora**: these are 'poetic plurals'; i.e. the poet, as often, uses plurals where we would expect singulars. This is often due to the demands of metre, but there is usually some idea of mass or plurality, and it frequently happens with parts of the body, which are liable to come in pairs. We must translate the words in the singular. A number of these have been encountered in the course of the passage and it would be prudent to make sure that the principle has been mastered.
ll.111–12: though nominative, **hoc** scans long in these lines. It should nevertheless be pronounced as a short vowel. See Allen, *Vox Latina*, pp.76–7.
ll.118–19: the enmity between Carthaginians and Romans which Dido had prayed for was to destroy her people in a way in which Anna could not anticipate. The Romans razed Carthage to the ground. Anna is presumably thinking of the defencelessness of Carthage against Pygmalion and the African tribes when the queen is dead. The educated reader would no doubt recall Andromache's lament for Hector in *Iliad* 24 (ll.725–45).
ll.120–1: the soul was contained in the last breath of a dying man; the closest kinsman would try to catch it.
l.122 **sēmianimem**: scan as if the first 'i' were a 'j' (a consonant).

Exercise 14.1

ll.1ff.: the composer Hector Berlioz who, in his great opera *The Trojans*, set Book 4 to music, writes memorably of his early response to Virgil: 'One day, I remember, I was disturbed from the outset of the lesson by the line:

At rēgīna gravī iamdūmdum saucia cūrā.

Somehow or other I struggled on till we came to the great turning-point of the drama. But when I reached the scene in which Dido expires on the funeral pyre, surrounded by the gifts and weapons of the perfidious Aeneas, and pours forth on the bed – "that bed with all its memories" – the bitter stream of her life-blood, and I had to pronounce the despairing utterances of the dying queen, "thrice raising herself upon her elbow, thrice falling back", to describe her wound and the disastrous love that convulsed her to the depth of her being, the cries of her sister and her nurse and her distracted women, and that agony so terrible that the gods themselves are moved to pity and send Iris to end it, my lips trembled and the words came with difficulty, indistinctly. At last, at the line

Quaesīvit caelō lūcem ingemuitque repertā

at that sublime image I was seized with a nervous shuddering and stopped dead. I could not have read another word.

It was one of the occasions when I was most sensible of my father's unfailing goodness. Seeing how confused and embarrassed I was by such an emotion, but pretending not to have noticed anything, he rose abruptly and shut the book. "That will do, my boy," he said, "I'm tired." I rushed away, out of sight of everybody, to indulge my Virgilian grief.'

l.2 **strīdit**: 'The word accurately expresses the whistling sound with which breath escapes from a pressed lung.' – J. W. Mackail

l.7 **difficilīs obitūs**: poetic plural.

l.8: tension is balanced with resolution.

l.9: Dido was not fated to die so early and – Virgil states the matter clearly – she did not deserve to die (**meritā nec morte**).

l.13ff.: the multicoloured Iris appears in beauty and brings release to Dido at last. Do your pupils feel that the Iris episode is the right way to end this book? What are their reasons for their reactions?

★ p. 161: this picture of Iris is from the *Cōdex Rōmānus* in the Vatican.

At the start of Book 5, Aeneas sees flames rising from Carthage. He does not know what they mean, but he and his men feel a sense of foreboding. In Book 6 he goes on a strange journey to the Underworld and there in the **lūgentēs campī**, where those whom love has destroyed dwell, he meets Dido:

Amongst them, with her death-wound still bleeding, through the deep wood
Was straying Phoenician Dido. Now when the Trojan leader
Found himself near her and knew that the form he glimpsed through the shadows
Was hers – as early in the month one sees, or imagines he sees,
Through a wrack of cloud the new moon rising and glimmering –
He shed some tears, and addressed her in tender, loving tones: –
Poor, unhappy Dido, so the message was true that came to me
Saying you'd put an end to your life with the sword and were dead?
Oh god! was it death I brought you, then? I swear by the stars,
By the powers above, by whatever is sacred in the Underworld,
It was not of my own will, Dido, I left your land.
Heaven's commands, which now force me to traverse the shades,
This sour and derelict region, this pit of darkness, drove me
Imperiously from your side. I did not, could not imagine
My going would ever bring such terrible agony on you.
Don't move away! Oh, let me see you a little longer!
To fly from me, when this is the last word fate allows us!
Thus did Aeneas speak, trying to soften the wild-eyed,
Passionate-hearted ghost, and brought the tears in his own eyes.
She would not turn to him; she kept her gaze on the ground,
And her countenance remained as stubborn to his appeal
As if it were carved from recalcitrant flint or a crag of marble.
At last she flung away, hating him still, and vanished
Into the shadowy wood where her first husband, Sychaeus,
Understands her unhappiness and gives her an equal love.
None the less did Aeneas, hard hit by her piteous fate,

Weep after her from afar, as she went, with tears of compassion.
(*Aeneid* 6, 450–76, translated by C. Day Lewis)

Chapters 15–17 Livy

With two exceptions, these chapters consist of excerpts from Livy, Book 21. We think that pupils will probably find Livy the most difficult of the writers in Part III and we have slimmed down the text substantially. Cuts apart, we have altered nothing that he wrote and many challenging passages remain. Livy's word order frequently poses problems and you must stress how vital it is to pay great attention to word endings. He makes much use of the historic present and frequently omits the verb 'to be'. The writing can be taut and telegraphic, both vivid and dramatic. At other times, his sentences come close to collapse. As L. P. Wilkinson writes, 'he is apt to tack on participles, letting one incident develop out of another as in real life; or to let a muddled sentence represent a confused scene.'

We refer readers to the admirable edition of Book 21 by P. G. Walsh, Bristol Classical Press, and to Sir Gavin de Beer's *Hannibal*, Thames and Hudson.

Chapter 15

The greatest war in history

The excerpts in this chapter are from Book 21, 1, 2, 4, 21, 27 & 28.

p.162, para. 2: Tacitus (*Annals* 4, 34) tells how Livy praised Pompey so enthusiastically that Augustus called him 'the Pompeian'. His republican sympathies led him to write of Cassius and Brutus, the assassins of Julius Caesar, with respect.

★ p. 162: this bust of Hannibal is in the British Museum.

l.1 **praefārī**: in addition to the Preface to the entire *History*, Livy wrote fresh introductions at suitable points. He wishes to stress the momentous nature of the Second Punic War (218–201 BC).
ll.4–5 **neque validiōrēs . . . hīs ipsīs**: here Livy consciously echoes Thucydides's preface to his history (1, 1). Livy hoped to put his history on a level with that of his famous Greek predecessor. He aimed to write as definitive a history of Rome as Thucydides had of the Peloponnesian War, and he wanted to emphasize that he was part of a great tradition of historiography.
ll.10–11 **īnferrent, crēderent**: subjunctives are regularly used to give a reported reason.

★ p. 163: the ruins of Carthage are not from the Phoenician city, which was almost completely obliterated by the Romans in 146 BC, but from a new town founded on the site by Julius Caesar.

l.12 **fāma**: the story appears in the work of Polybius (200–after 118 BC), the Greek historian whose work dealt with the rise of Rome to world power. He was one of Livy's main sources for the part of his history dealing with Hannibal. He in fact crossed the Alps in Hannibal's footsteps.

The episode Livy describes took place in 237 BC when Hannibal (247–183/2) was nine.
l.17: Hamilcar died, by drowning, in 229/8 BC. Hasdrubal was in command in Spain until he was murdered in 221.
ll.25ff.: the character of Hannibal. Livy's model here is Sallust's portrait of Catiline (*Catiline* 14–16). It is a fine piece of writing: clauses and phrases are constantly balanced and alliteration (e.g. **perfidia plūs quam Pūnica**, l.40) and the omission of verbs give the passage animation.

Summarizing Livy's character sketch of Hannibal, de Beer writes, 'Supreme at fighting, with a total disregard of danger, it is not surprising that all these qualities endeared him to his men to a fanatical extent.' According to Cornelius Nepos, Hannibal was a scholar as well as a soldier. There is no reason to believe that he was particularly treacherous or cruel (l.40). As for piety (**nihil sānctī**, l.40), an irreligious man does not travel more than 700 miles to worship his guardian diety as Hannibal did in 218 BC. Livy's description here reflects Roman bias.

★ p. 165: in the foreground of this picture of Saguntum is the Roman theatre (built later). The citadel dominates the site.

l.44: we are now in January 218 BC.
ll.60ff.: Hannibal's crossing probably took place at Fourques, opposite Arles. It was a sensational achievement. He crossed a river

almost 1 km. wide with 50,000 infantry, 9,000 cavalry and 37 elephants.

★p. 167: opposite Avignon, the Ile de Bartelasse divides the Rhone into two swiftly flowing streams. It may have been here that Hanno crossed over his light troops.

l.70 **ratibus iūnctīs**: the meaning given in the gloss seems the likely one, but the natural sense of the phrase suggests that they joined the rafts into larger units for the crossing, or made bridges of them.

Exercise 15.2

We have added the word **Hannibal** to Livy's Latin to make the meaning clearer.
ll.3ff.: Livy added this description of the Gauls to what he found in Polybius. His reading of Caesar may have inspired this lively passage.
l.15: Hannibal's contempt for the Gauls in this encounter led him never to put too much reliance on them later when they were his allies against Rome. There is a hint in the Latin that, while the Romans made heavy weather of the **Gallicōs tumultūs** (which could mean 'the turbulent Gauls') Hannibal easily managed to cut them down to size. (This note is relevant to Question 4.)
Chapter 28 of Book 21 contains a memorable account of the crossing of the elephants. It would be worthwhile reading this to your pupils in translation (e.g. Aubrey de Sélincourt's, Penguin).

Chapter 16

Hannibal reaches the Alps

Book 21, 32.6–35.3

ll.2–6: this is one of those passages where Livy shows his unsurpassed historical empathy. He enters into the minds of the Carthaginians as they gaze in appalled terror at the sight which confronts them. Do you feel that this should be thought of as imaginative writing, or could one apply to Livy here Hobbes's remark on Thucydides, that he did not 'enter into men's hearts further than the acts themselves guide him'? This was the situation; this *must* have been the reaction.
l.7ff.: this episode may have taken place in the Col de Grimone.
ll.22–3 **plūribus . . . factīs**: Hannibal is trying to give the impression that the whole army is in the camp. Thus the enemy will not suspect the existence of a storming party.
l.28ff.: the changing reactions of the Gauls are vividly conveyed.
l.56 **populum**: might mean 'region' or 'district' here.
l.57 **suīs artibus, fraude et īnsidiīs**: 'Pūnica fidēs' was an ironical expression referring to the proverbial treachery of the Carthaginians. Again Roman prejudice makes itself felt.
l.69: Hannibal may now be in the gorge of the River Guil.

Exercise 16.2

A short passage for translation in place of the usual comprehension exercise.

★p. 179: a Carthaginian silver coin from Hannibal's time shows an African elephant. Comparing the size of the driver with the animal, one can see how small the African elephant was. (British Museum)

Chapter 17

Hannibal crosses the Alps

Book 21, 35.4–37, Book 39, 51, Juvenal 10, 147–67 (with one line omitted).

ll.1–3: it looks as if Hannibal was simply hoping for the best in his search for a pass over the Alps.
ll.7ff.: another poignant passage in which Livy shows his empathy in his narrative.
l.11: this may have been near the Col de la Traversette, which is almost 10,000 feet high. John Ball wrote of this col: 'The view suddenly unfolded at the summit, extending in clear weather across the entire plain of Piedmont as far as Milan, is extremely striking.'
l.16 **in manū ac potestāte**: a legal expression, suggesting Hannibal's belief that he and his men will soon be the lawful masters of Italy and Rome. Note the movement in this paragraph from abject despair to superb self-confidence.
l.17 **nē . . . quidem** is mystifying. It seems to be there to contrast with **cēterum** (l.18). The enemy did not bother them, but the terrain did.
p.182: this is an exciting but difficult passage in the Latin. We have therefore translated it.
l.34ff.: this story has been ridiculed by many

writers, but there is no reason to doubt it. De Beer cites a successful experiment by the Director of the British Museum in 1956. 'Blocks of limestone were covered with logs and were kindled into a hot fire, after which cold water containing ten per cent of acetic acid was thrown on to the stones. There was a cloud of steam and much hissing as the acid attacked the limestone, and when the cloud had cleared the limestone blocks were found to have been split.' De Beer concludes with a neat rebuke to the doubters: 'It is never safe to doubt factual statements of a practical nature by serious classical authors.'

ll.42–3: the elephants suffered particularly from the lack of food.

l.47–8 **locīs molliōribus et accolārum ingeniīs**: literally, 'the places (being) softer, i.e. easier to traverse, and the natures of the inhabitants (being softer, i.e. gentler)'.

l.49: Hannibal got across with 12,000 African infantry, 8,000 Spaniards and all of the 37 elephants. He told a Roman prisoner that he had lost 36,000 men since crossing the Rhone.

★ p. 183: Monte Viso is, at 3,841 metres, the highest peak of the Cottian Alps. It looms over the Col de la Traversette, by which Hannibal may have crossed.

Indicative and subjunctive in subordinate clauses

We here summarize the use of the subjunctive in various kinds of subordinate clauses. By emphasizing the similar ways in which the subjunctive is used in these clauses, we hope that we shall have saved much labour. However, you may wish to add more practice sentences to our exercise.

Exercise 17.3

l.1 **T.** = Titus. Explain that Roman names are usually given in full in translation. There is a list of names on p.188.

ll.19–20: cf. Thomas North's translation of Plutarch: '. . . Titus Livius, that famous Historiographer, writeth, that Annibal called for the poyson he had ready for such a mischiefe, and that holding this deadly drinke in his hand, before he dranke he sayd: Come on, let us rid the Romans of this payne and care, sith their spight and malice is so great, to hasten the death of a poore old man that is halfe dead already.'

ll.23–4: Hannibal died when he was sixty-four. He was cornered, in the words of Plutarch, 'like a bird which had grown too old to fly and had lost its tail feathers'.

Extract from Juvenal

l.4: as we have seen, Hannibal commanded the Carthaginian troops in Spain from 221 BC; he crossed the Pyrenees and then the Alps in 218 BC.

l.5 **trānsilit**: a striking use of enjambement (cf. **ānulus** (l.19)). Hannibal could move as swiftly as Alexander or Caesar.

l.8 **āctum . . . nīl est**: Lucan wrote of Caesar, 'nīl āctum crēdēns cum quid superesset agendum'.

l.19: it was particularly appropriate that the poison from a ring should take vengeance for Cannae. The messenger who took the news of the battle to Carthage confirmed his story by pouring out before the senators a vast heap of gold rings taken from the hands of the Roman knights who had been killed.

Samuel Johnson (1709–84) wrote a famous version of Juvenal's tenth satire, adapting it to the career of Charles XII of Sweden (1682–1711):

On what foundation stands the warrior's pride,
How just his hopes let Swedish Charles decide;
A frame of adamant, a soul of fire,
No dangers fright him, and no labours tire;
O'er love, o'er fear extends his wide domain,
Unconquer'd lord of pleasure and of pain;
No joys to him pacific scepters yield,
War sounds the trump, he rushes to the field;
Behold surrounding kings their pow'r combine,
And one capitulate, and one resign;
Peace courts his hand, but spreads her charms in vain;
'Think nothing gain'd,' he cries, 'till nought remain,
On Moscow's walls till Gothic standards fly,
And all be mine beneath the polar sky.'
The march begins in military state,
And nations on his eye suspended wait;
Stern famine guards the solitary coast,

And winter barricades the realms of Frost;
He comes, not want and cold his course delay;
Hide, blushing glory, hide Pultowa's day:
The vanquished hero leaves his distant bands,
And shews his miseries in distant lands;
Condemn'd a needy supplicant to wait,
While ladies interpose, and slaves debate.
But did not Chance at length her error mend?
Did no subverted empire mark his end?
Did rival monarchs give the fatal wound?
Or hostile millions press him to the ground?
His fall was destin'd to a barren strand,
A petty fortress, and a dubious hand;
He left the name, at which the world grew pale,
To point a moral, or adorn a tale.
(*The Vanity of Human Wishes,* 191–222)

For pupils:
Questions and activities: Is Johnson's adaptation more or less scornful than the original? Discuss the lasting relevance of the poets' theme. Do you take an equally jaundiced view of the achievements of great generals?

Look up details of Napoleon's crossing of the Alps in 1800. In Jacques-Louis David's great picture 'Napoleon crossing the St. Bernard Pass', Hannibal's name is carved on a rock.

⋆p. 188 this bronze steelyard balance is from Pompeii. The fulcrum is eccentric; the scale pan hangs from the shorter arm and the counterweight hangs from a loop which is free to move along a graduated scale on the longer arm of the fulcrum.

⋆p. 189 This relief of an elephant is from a sarcophagus dating from *c.* AD 130–150. Bacchic revellers ride on top of the elephant. (Fitzwilliam Museum, Cambridge)

Chapters 18–20 Ovid

The *Amōrēs* and the *Ars Amātōria* are to be found in E. J. Kenney's Oxford Classical Text of Ovid's love poetry. *Amōrēs* 1 has been edited by John Barsby, Oxford, and *Ars Amātōria* 1 and *Metamorphōsēs* 8 by A. S. Hollis, Oxford. For English-speaking readers, the *Trīstia* are available only in the Loeb edition (A. L. Wheeler, Heinemann).

All the love poetry and the poems from exile have been translated by Peter Green in two valuable Penguin Classics (*Ovid: The Erotic Poems* and *Ovid: The Poems of Exile*). Guy Lee's translation of the *Amores* (John Murray) Bis highly recommended.

Chapter 18

Ovid tells the story of his life

Trīstia 4, 10: 1–6, 9–12, 15–42, 45–6, 49–52, 55–60, 65–8, 93–4, 97–8; 1,3: 1–22, 79–82, 85–9; *Amōrēs* 2, 6: 1–4, 11–14, 17–20, 37–8, 45, 48

Sulmo: Ovid's love for the place of his birth is evident in his poetry, where its lushness and fertility are expressively conveyed. Modern Sulmona repays the compliment. Its official documents are headed with the initials SMPE (**Sulmo mihī patria est**, l.3) on the analogy of SPQR at Rome.

l.1 **tenerōrum lūsor amōrum**: Ovid draws attention both to the tenderness and to the playfulness of his love poetry
l.2 **nōrīs** (= **nōverīs**): perfect subjunctive in a purpose clause. The perfect subjunctive is used instead of present since, while **nōscō** = I get to know, **nōvī** (perfect) = I know.
l.4: the distance is in fact 94 Roman miles or 86 English miles.
l.8 **tribus . . . quater**: Ovid's roundabout way with numbers is partly due to the requirements of metre. It is impossible to use **duodecim** in Latin verse.
l.9 **Lūcifer**: the light-bringer, i.e. the morning star.
l.12 **īnsignēs . . . virōs**: it is not impossible that, when Ovid and his brother came to Rome in about 31 BC, they were taught by Horace's teacher, the **grammaticus** Orbilius. Ovid's teachers of rhetoric were Arellius Fuscus and Porcius Latro. The boys' father wanted to give both of them the education needed for the practice of law. This, he hoped, would prove the basis for successful political careers.
l.16 **Mūsa**: cf. l.46. The nine Muses were born in Pieria at the foot of Mount Olympus and dwelt on Mount Helicon in Boeotia. Cf. Michael Drayton (1562–1631) on Ben Jonson:

Next these, learn'd Jonson, in this list I bring,
Who had drunk deep of the Pierian spring.

★ p.192: this mosaic floor from a villa at Trier dates from the second century AD and shows the nine Muses: Top row: Thalia, Muse of Comedy, wreathed with ivy and holding a shepherd's crook and a comic mask; Terpsichore, Muse of the Dance, with a lyre; Clio, Muse of History, with a scroll; Middle row: Euterpe, Muse of Lyric Poetry; an unidentified Muse; Erato, Muse of Love Poetry, holding a lyre; Bottom row: Urania, Muse of Astronomy, with a celestial globe; two unidentified Muses. The unidentified ones must be Calliope (Epic), Melpomene (Tragedy), and Polymnia (Sacred Song). (Rheinisches Landesmuseum, Trier)

l.18: Maeonia is the Western district of Asia Minor where tradition maintains that Homer was born.
ll.21–2: there is a story that Ovid's father rebuked him as a boy for scribbling poetry instead of doing his homework. The boy cried, 'Parce mihī! nunquam versificābo, pater!' (Forgive me, father! I'll never write a verse!). This is a complete pentameter. (The **-o** at the end of **versificābo** scans short.) Pope's imitation of ll.21–2 is famous:

> As yet a child, nor yet a fool to fame,
> I lisped in numbers, for the numbers came.
> *Epistle to Dr Arbuthnot,* 127–8

l.2: the **equitēs illūstrēs** (who had a capital reserve of at least 400,000 sesterces – the minimum required for entry to the Senate) could wear two broad purple stripes on their toga. Other **equitēs** wore a narrow purple stripe and a golden ring. Augustus himself may have wished to support the political advancement of Ovid's family. See R. Syme: *The Roman Revolution*, pp.358–9.
l.28 **perit**: historic present, conveying an immediacy of feeling – especially after the elegant circumlocution of the previous line.
ll.29–30: he could not take the first step of a senatorial career until the age of thirty (when he could assume the quaestorship).

★ p. 193: this statue dates from the first century AD.

l.31 **cūria**: i.e. a senatorial career.
l.36 **ōtia**: Ovid does not mean that he did nothing. He was in fact a highly productive poet. He is referring to a life removed from public affairs. **ōtium** frequently has this meaning.
l.39: Propertius (*c*.50-*c*.16 BC) addressed his love poetry to a woman whom he calls Cynthia.
l.41: this line could mean either that Ovid heard Horace recite, or that Horace's poetry charmed his ear, or both. **numerōsus**: as we have seen in Part II, Horace (65–8 BC) made use of many metres in his *Odes*. No major Latin poet chose to follow in Horace's footsteps in his use of Greek lyric metres.
l.42 **culta**: the word means something like 'elegant' or 'polished'. Augustan poets were not warblers of 'native wood-notes wild'. They aimed at a highly-wrought sophistication.
ll.43–4: Virgil (70–19 BC); Tibullus (60/55 – 19 BC).

★ p. 194: This contorniate-portrait of Horace dates from the fourth century AD and is in the British Museum.

★ p. 194: this bronze statuette of Apollo, from the House of the Red Walls in Pompeii, is a mere 27 centimetres high. The god's mantle is draped over the column on which he leans; his lyre has silver strings.

l.47: for public recitals of poetry, see Pliny, *Letter* 47: 'This year has produced a great crop of poets; throughout the month of April there has scarcely been a day when nobody has given a recital. I am delighted that literature is flourishing and that so many talented men are coming forward to display their art; but the audience response is depressingly unenthusiastic. Most people sit in the porticos outside and while away the time they should be spending listening in idle chatter . . . At long last – and even then slowly and reluctantly – they go in, but they do not stay to the end; they go out before then, some of them slipping away inconspicuously, but others leaving quite brazenly.'
l.50: poets would choose a 'poetic' name for the object of their passion; it would be of the same metrical quantity as the real name (thus Clodia = Lesbia).
l.54: Ovid claims elsewhere (*Trīstia* 2, 353–4) that, while his poetry was risqué, his life was pure:

> crēde mihī, mōrēs distant ā carmine nostrō –
> vīta verēcunda est, Mūsa iocōsa mea.

ll.57–8: Ovid was on Elba when the news of his banishment came. He was not in fact condemned to **exsilium** but to the less severe

sentence of **relēgātiō**. This allowed him to keep his property and rights as a citizen. From Elba he went to Rome to prepare for his departure.
ll.59–88: L. P. Wilkinson writes of the poem from which these lines are taken, 'It is a sincere and vivid record of a poignant personal experience, a thing rare in ancient poetry, except on the subject of love or death, though Cicero's Letters offer us counterparts in prose.'
l.69: Jupiter wielded the thunderbolt (which combined thunder and lightning).
l.75: this was Ovid's third wife. His daughter was the child of his second wife. We do not know her name.
l.86: her **pietās**, i.e. her dutiful love of her husband, is as imperious as Caesar in its demands.

★p. 196: Augustus is here represented as a pious citizen, performing a sacrifice or attending a religious ceremony, with part of the toga drawn up to veil his head. (Museo delle Terme, Rome)

Grammar: 'cum'

The uses of **cum** are very complex and we have simplified them. We are confident that, despite a certain amount of glossing over, we provide a more than adequate summary for the GCSE examination.

Exercise 18.3

Amōrēs, 2, 6, brutally reduced.
The poem is a parody of the commemorative dirge. First the mourners are addressed (1–6), then there is the lamentation (7–14), followed by an account of the bird's death (15–16). There were good precedents for poems commemorating a dead pet. The most famous is Catullus 3 (see p.112).

The humour implicit in the personification of the birds is beautifully handled. A talking bird is, of course, an especially apt candidate for such treatment. Corinna's parrot is a very special bird, an exotic pet, a lover's gift and a loving creature. Is there a measure of genuine pathos alongside the humour?
ll.3–4: at a Roman funeral, professional female mourners were hired to pull out their hair, beat their breasts and tear their cheeks with their fingernails.

★p. 200: this mosaic is from Pompeii. It dates from the first century AD.

★p. 201: this mosaic of doves drinking from a bowl is from Hadrian's Villa at Tivoli (second century AD). It is probably a Roman copy of an original by Sosias of Pergamon. (Capitoline Museum, Rome)

Chapter 19

Ovid the lover

Amōrēs 1, 5 (omitting ll.11–12), *Ars Amātōria* 1, 89–102, 107–10, 113–24, 127–32; *Amores* 3, 2: 1–14, 19–24, 65–84; *Metamorphōsēs* 8, 679–94.

Extract 1

p. 202 *Siesta time*: the title is taken from Guy Lee. This is the first poem in Latin literature simply describing love-making. It is in no way a prurient piece. The mood is one of relaxed hedonism.
l.3: the windows are wooden shutters with transverse slats; one shutter is open.
l.7 **verēcundīs**: the girl who arrives is hardly shy!

★p. 202: this superb statue of Parian marble is a Roman replica of a Hellenistic original, derived from the Cnidian Aphrodite of Praxiteles which was modelled on a famous courtesan. Our statue was found in the seventeenth century in a house near San Vitale, Rome, and is called the Capitoline Venus because it stands in the Capitoline Museum in Rome.

Extract 2

p.204, l.1: Ovid frequently refers to the theatre as a promising location for meeting girls.
ll.5–8: the similes of the ant and bees are taken from Virgil (see pp.152 and 132). The use of similes from high (and patriotic) epic in such a context calls for discussion.

Their Virgilian echoes apart, these are characteristically Ovidian similes, playful and not a little absurd as the humble insects are called upon to evoke the rush of the female *beau monde* (**cultissima fēmina**, l.9) to the theatre.

As the poem progresses, you may like to discuss with your pupils whether it it is right to view it as subversive of Augustan values.
ll.11–12 Dryden wittily translates:

To see, and to be seen, in Heaps they run;
Some to undo, and some to be undone.

★ p.204: the theatre at Orange dates from the first century AD. The massive back wall of the stage, with its turret-like projections, was originally decorated elaborately with marble architecture.

l.13ff.: Ovid in this passage satirizes the aetiological tendency of Hellenistic poetry (i.e. its proneness to go back to the causes of things); and he pokes fun at the way in which Augustan literature makes use of early Roman legend to add grandeur and dignity to the nation's heritage (see especially Livy l, 9 and Virgil, *Aeneid* 8, 635–6). Augustus had at one time considered adopting the name of Romulus, as the second founder of Rome. It is an amusingly impudent idea to present him as the patron of pick ups.

ll.15–16: Ovid stresses the primitive conditions of the theatre in the early days of Rome. The seats were not of marble and there were no awnings stretched over the top, as there were in the Theatre of Pompey in his time. Augustus claimed that he had found Rome brick and left it marble; but he would not have altogether appreciated Ovid's oblique tribute to Augustan theatres, with its implication that splendid surroundings are the perfect location for seduction.
l.17: in 18 BC Augustus had legislated that women should occupy only the back rows in the theatre. He aimed to keep the sexes apart. Ovid hints at the ineffectiveness of such legislation.

★ p.206: this relief was put on the Basilica Aemilia in the Forum on its restoration, financed by Augustus, probably in 14 BC.

l.31 **sī qua repugnārat nimium**: 'a notion which recurs time and again in the love-poets. It was right and proper for the girls to put up a show of reluctance, but not to carry their opposition too far.' (A. S. Hollis)
ll.35–6: the return to Romulus is a good example of a standard narrative technique, i.e. ring-composition (it was Romulus who started it all – this is how it happened – how splendid of Romulus!). The joke on **commoda** is precisely the sort of crack that may have got under Augustus's skin and contributed to Ovid's downfall. Augustus did in fact have difficulty in making the terms of service in the Roman army attractive. In effect Ovid is saying, 'If they could offer a pretty girl as a side-attraction nowadays, that would solve the recruiting problem.' (Hollis) Dryden translates:

Thus *Romulus* became so popular;
This was the Way to thrive in Peace and War;
To pay his Army, and fresh Whores to bring;
Who wou'd not fight for such a gracious King!

Extract 3

l.1: Ovid is at the races in the stadium of the Circus Maximum in Rome. This huge rectangle with semicircular ends was surrounded by tiers of covered seats which held 250,000 spectators.

★ p.207: the Circus Maximus lay in the Vallis Murcia, between the Palatine and Aventine Hills.

l.9: modern equivalents of **carcerēs** can be seen on today's race-courses; the **carcer** is **sacer** because it is a kind of gateway and all gateways and the like belonged to the god Janus. The chariots rushed from the starting boxes, beginning anticlockwise on the first of what were usually seven laps of the 550 metre track. They dashed along the right on a central division (**spīna** backbone) in the stadium. Each time a lap was finished, an object made of marble, either in the shape of an egg or a dolphin, was removed from the **spīna**.

★ p.208: this statue of the praetor dropping the white cloth to start the race is from the Capitoline Museum in Rome.

★ p.208: this terracotta relief made in Italy in the first century AD shows a chariot race. A four-horsed chariot approaches the turning-post with its three decorated columns. The driver reins in his horses. A **jūbilātor**, a rider who encourages the contestants, has just turned. By the turning-post, there crouches – just visible – a fallen charioteer.

ll.29–30: races could in fact be stopped and re-started in the manner described here.
After reading these poems set at the theatre and the races, it may be profitable to discuss with your pupils why Ovid has chosen the

background of entertainment and sport for his love poetry.

Numbers

p.211: at present these distributive numerals and numeral adverbs are on a GCSE syllabus (MEG) and we have felt obliged to include them. However, we suggest that you omit them, ensuring only that **semel**, **bis** and **ter** are known.

Exercise 19.1: Baucis and Philemon

(*Metamorphōsēs*, 8, 611–724)

No telling of this story exists in Latin literature before Ovid. However, the theme of a god or a great figure entering a humble house was a popular one. In *Aeneid* 8, for example, Evander invites Aeneas to follow the example of the demigod Hercules and stoop to enter his simple dwelling (364–5).

The range of feeling and tone in the *Metamorphōsēs* is vast. In this episode, Baucis (the wife) and Philemon (the husband) show an ideal simplicity, looking back to Rome's rustic beginnings (cf. Horace's story of the town and country mouse (*Satires* 2, 6, 79–117)). There is little of Virgil's intensity or depth of emotion. But Ovid, deftly combining lightness of touch with an underlying seriousness, creates an episode in which we willingly suspend our disbelief. No poet, except for Homer in the *Odyssey*, has brought the miraculous into the real world more convincingly.

Ovid conveys the key Augustan virtue of **pietās** with panache and humour. The contrast with the grave profundity of Virgil's treatment of the same concept is fascinating. Ovid's use of detail is particularly effective. 'I see *Baucis* and *Philemon* as perfectly before me,' wrote Dryden, 'as if some ancient Painter had drawn them.'

Here is Dryden's version of part of this passage (ll.6–10):

> One Goose they had ('twas all they cou'd allow)
> A wakeful Cent'ry, and on Duty now,
> Whom to the Gods for Sacrifice they vow:
> Her, with malicious Zeal, the Couple view'd;
> She ran for Life, and limping they pursu'd:
> Full well the Fowl perceiv'd their bad intent,
> And wou'd not make her Masters Compliment;
> But persecuted, to the Pow'rs she flies,
> And close between the Legs of *Jove* she lies:
> He with a gracious Ear the Suppliant heard,
> And sav'd her Life.

ll.8–10: 'Their vain attempts to catch the goose make a contrast with the unhurried and dignified gods, and also prevent the narrative from becoming too elevated.' (Hollis) The goose, as Dryden perceives, becomes a suppliant, running to the gods to beg their assistance. Supplication was a serious business in the ancient world, and the god or human who was called upon for help would feel a certain obligation to give it. The gods do so here.

★ p.213: this statue of a child playing with a goose is a Roman copy of a Hellenistic composition of the third century BC. (Louvre Museum, Paris).

Chapter 20

Ovid in exile

Trīstia 1, 2: 1–2, 19–26, 31–4; 3, 10: 1–10, 13–14, 17–22, 25–6, 31–2, 37–8, 47–50, 53–8, 61–70, 75–8; 3, 3: 1–4, 7–18; 5, 7: 39–46, 51–6; *Amōrēs* 3, 9: 1–6, 35–40, 59–62, 65–8.

The Emperor had insisted that Ovid depart in mid-winter, thus increasing the likelihood of a stormy voyage. Tomis, now Costantza in Rumania, is near the mouth of the Danube. In modern times, Costantza is a popular holiday resort for Eastern Europeans with a temperature ranging from 80°F in the summer to 25°F in the winter. Inland, however, it sometimes drops to 20° or more below 0°F.

One result of Ovid's banishment was that this famous love poet totally lost interest in sex (*Ex Pontō* 4, 2, 33–4).

l.6 **Tartara**: Tartarus was the place of punishment in the Underworld. Hesiod writes that it is as far below the earth as heaven is above.

l.8 **hic . . . ille** the latter . . . the former. Explain this usage to your pupils: **hic** (this) means 'the latter' because of the two items previously mentioned it refers to one closer to it.

l.21: Hister usually refers to the lower part of the Danube.

l.37 **pontum**: **pontus** means 'sea'; but it can refer specifically to the Black Sea.

★p.220: a Scythian bronze dating from about 500 BC. cf.★ III, p.117. (British Museum)

★p.220: this portrait of a woman from Roman Egypt dates from the first half of the second century AD. It is a mummy-portrait from Antinoopolis and is in the Louvre Museum in Paris.

ll.87–8: barbarians are those who cannot speak Greek (they say 'barbar'). In fact, as Ovid says, a debased Greek was spoken in the area.
ll.89–90 **ē mediō . . . verba**: words in common use as opposed to literary or educated language.

Elsewhere Ovid complains that writing poetry which he can read to nobody is like dancing in the dark. 'If you put Homer himself in this land, believe me, he would become a Getan.' (*Ex Pontō* 4, 2, 33–4, 21–2)
Ovid learnt the local languages and received kind treatment from the people of Tomis, who exempted him from taxes. A statue of Ovid stands in the main square of Constantza (see illustration on p.217).

Exercise 20.2

Ovid's fellow-poet Tibullus wrote about love and the countryside. His elegiac couplets are plain but polished. He is never strident but conveys the depth of his feelings with a quiet intensity. We cannot be sure how old Tibullus was when he died in 19 BC (in the same year as Virgil). Ovid later referred to him as dying young:

tē quoque Vergiliō comitem nōn aequa, Tibulle,
mors iuvenem campōs mīsit ad Elysiōs.

This poem is a funeral lament (compare and contrast the elegy for Corinna's parrot (p.200)). The chief mourner is addressed (1–6); there follow the lament (7–12), the **cōnsōlātiō** (13–18) and a brief **requiem** in conclusion.

There is dignity and sincere emotion in this poem, as well as a high seriousness in its tragic vision of a world where death seems at first to be the only reality. But the **cōnsōlātiō**, with its delightful evocation of a poets' heaven, movingly asserts the victory of poetry over death and leads to the poignant diminuendo of the concluding prayer.

The fine simplicity of these lines recalls Catullus's lament for his brother (p.126).
ll.1–4: the word 'elegy' is more likely to be connected with some foreign word meaning a flute – so that an elegy is a flute song. Early elegies are in fact not usually laments.
ll.13–18: this beautifully poised passage no doubt owes something to Socrates's speculations about death at the end of Plato's *Apology*. It inspired Thomas Dekker to set a fraternity of dead Elizabethan writers in the Elysian fields (1607):

> Marlowe, Greene, and Peele had got under the shade of a large vine, laughing to see Nashe (that was but new come to their college) still haunted with the sharp and satirical spirit that followed him here on earth.

l.16 **docte**: in our Catullus chapters, we have concentrated on his more passionate and direct poetry. Nevertheless we hope that his dedication to his craft as a poet – and his mastery of it – will have come across. He is one of the most technically proficient of Latin poets.
l.18 **culte**: see n. on l.42 in Chapter 18 (Teacher's Book, p.96).

★p.224: this glass urn, containing the ashes of a cremated body, dates from the first or second century AD. (Harrow School)

★p.224: in this relief of a funeral from a Roman sarcophagus of the first century BC or AD, musicians lead the way; the dead man reclines as if he is observing his own funeral; his family follow behind. (Museo Aquilano, Rome)

THE METRES OF THE POEMS OF PARTS II AND III

1 TEACHING SCANSION

Latin poetry was written to be read aloud. Its publication was usually at a **recitātiō** and the ancients attached prime importance to the sound of verse. The purpose of teaching scansion is simply to help your pupils to read rhythmically and any lesson on metre should start with one or more readings aloud. If from the start your pupils have learnt correct pronunciation, they may well feel the rhythm of metres they have never studied, before they know anything about scansion. It would be wise to start the study of metre not with dactylic hexameters or elegiac couplets, but with a very simple metre such as iambics; Horace, *Epode* 2 (Part II, p.121) or Catullus 4 (Part III, p.120) would provide ideal points of departure.

2 QUANTITY

The scansion of Latin verse of the classical period is quantitative, not as in English accentual. Syllables are either light or heavy, regardless of where the accent falls on any given word.
Heavy syllables:
(a) All syllables are heavy which contain a long vowel or a diphthong, e.g. **lāētī, sōlēs, Rōmānī** (for the purposes of scansion heavy syllables are marked with a macron –, light syllables with the symbol ˘; this convention sometimes results in syllables containing a short vowel being marked with a macron –; see below).
(b) If a short vowel is followed by two consonants, whether in the same or different words, the syllable is heavy, e.g.

tā̲ntāē | mōlĭs ĕ|rā̲t Rō|mānā̲m|cōndĕrĕ|gēntĕm.

The syllables underlined are heavy although in each case the vowels are short.
(c) Exceptions to rule (b): if a short vowel is followed by a combination of mute (**p,t,c,b,d,g**) and liquid **r** (and more rarely **l**), the syllable may be either light or heavy, e.g. **pā̆tris, volū̆cris, latē̆brae**. This is really a question of pronunciation; such syllables can either be pronounced **pāt-ris** or **pă-tris** (**tr** making one sound).

3 ELISION

A final open vowel followed by a vowel at the beginning of the next word is elided, as in the French *c'est*, but in Latin the elision is not written, e.g.

crēdō equidem (**ō** elides before following **e**)
multa quoque et (**e** elides before following **e**).

More surprisingly a final syllable ending in **-m** is elided if the next word starts with a vowel, e.g.
audiam et haec (**am** elides before following **e**)
īre iterum in lacrimās(**e** elides before following **i** and **um** elides before following **i**).
In reading Latin verse the elided vowel or syllable should be lightly sounded.

4 THE METRES

1 Iambics

An iambic metron contains two iambic feet:
⏑ – ⏑ –.
Normally a spondee (– –) may be substituted for an iamb in the first foot of each metron:
⏓ = ⏑ –
The last syllable in the line in all types of metre may be long or short.

The commonest iambic line is the iambic trimeter; this scans, counting in feet, as follows:

1 ⏓ = | 2 ⏑ – | 3 ⏓ ∧ = | 4 ⏑ ∧ – | 5 ⏓ = | 6 ⏑ ⏒

There is a caesura, i.e. a rhythmical pause between words, half way through the third or fourth foot, marked ∧.

The metre of Horace, *Epode* 2 (Part II, p.121) is an iambic trimeter followed by an iambic dimeter:

1 bĕā|2 tŭs īl|lĕ ∧ 3 quī | 4 prŏcūl | 5 nĕgō|6 tĭīs

1 ūt prīs|2 că gēns | 3 mōrtāl|4 ĭŭm

Catullus 4 (Part III, p.120) is written entirely in pure iambics:

phăsē|lŭs īl|lĕ ∧ quēm | vĭdēt|ĭs, hōs|pĭtēs,
ăīt | fŭīs|sĕ ∧ nā|vĭūm | cĕlēr|rĭmŭs.

Limping iambics (Catullus 8, p.108 and 31, p.92)
These scan like iambic trimeters except for the last foot, which is always a spondee or – ⏑. This has a peculiar rhythmical effect, making the line drag or limp at the end.

mĭsēr | Cătūl|lĕ, dē|sĭnās | ĭnēp|tīrĕ
ēt quōd | vĭdēs | pĕrīs|sĕ pēr|dĭtūm | dūcās

2 Dactylic hexameters and elegiac couplets

The dactylic hexameter consists of six dactylic metra (– ⏑ ⏑); a spondee (– –) may be substituted for a dactyl in any of the first four feet; the fifth foot is nearly always a dactyl and the sixth is always a spondee or – ⏑.
There is usually a caesura in the middle of the third foot:

1 2 3 4 5 6
= ⏔ | = ⏔ | = ∧ ⏔ | = ⏔ | – ⏑ ⏑ | – ⏓

(if there is a weak caesura in the third foot, there are usually strong caesuras in the second and fourth)

ūrbs ān|tīquă fŭ|īt ∧ (Tўrĭ|ī tĕnŭ|ērĕ cŏ|lōnī) . . .
3rd foot strong caesura

quām Iū|nō fēr|tūr ∧ tēr|rīs măgĭs | ōmnĭbŭs | ūnăm
3rd foot strong caesura

pōsthăbĭ|tā cŏlŭ|īssĕ ∧ Să|mō. hīc | īllĭŭs | ārmă, . . .
3rd foot weak caesura

This is the metre used by Homer and all subsequent epic poets; it is used by Virgil in *Eclogues, Georgics* and *Aeneid*; by Ovid in the *Metamorphōsēs*; by Horace in the *Satires* and *Epistles*.
Elegiac couplets consist of a dactylic hexameter followed by the first half of the same (a hemiepes) repeated:

= ⏔ | = ⏔ | = ∧ ⏔ | = ⏔ | – ⏑ ⏑ | – ⏓
= ⏔ | = ⏔ | – ∧ – ⏑ ⏑ | – ⏑ ⏑ | ⏓

nīl nĭmĭ|ūm stŭdĕ|ō, Cāe|sār, tĭbĭ | vēllĕ plă|cērĕ,
nēc scīre | ūtrūm | sīs ∧ ālbŭs ăn | ātĕr hŏ|mō.

Elegiac couplets were used for epigrams early in the Greek tradition and soon developed into longer poems; they were the first Greek metre to be used in Latin verse. Catullus was the first Roman poet to use them for longer poems (e.g. 76 and 68 – the latter 160 lines long) besides epigram. They were Ovid's favourite metre; he uses them in the *Amōrēs, Ars Amātōria, Trīstia* etc.
Horace *Odes* 4, 7, **diffūgēre nivēs**, consists of a dactylic hexameter followed by the first half of a pentameter.

3 Lyric metres

There is a wide variety of lyric metres which first appear in the poems of Alcaeus and Sappho, who wrote in the Aeolic dialect about 600 BC (hence Horace's claim: 'dīcar . . . prīnceps Aeolium carmen ad Italōs dēdūxisse modōs.')
In these metres, which are dance rhythms, we cannot speak of feet; the unit is the line and many systems are constructed in four line stanzas. Most Aeolic metres contain a unit called the choriamb (– ⏑ ⏑ –) and the lines are built up round one or more of these; they consist of a base, choriamb(s), clausula (a closing cadence, which may take various forms, e.g. ⏑ ⏓, ⏑ – ⏓, ⏓). So in the passage quoted above:

prīncēps | Aēŏlĭūm | cārmĕn ăd Ī|tălōs

the line consists of base (– –), two choriambs (– ⏑ ⏑ – – ⏑ ⏑ –) and clausula (⏑ –).
The lyric metres illustrated in our selection are:

(a) Asclepiads

The Asclepiad metres all consist of base (– –), one or more choriambs, clausula.
Horace, *Odes* 3, 13 **ō fōns Bandusiae** (Part II, p.207);

ō fōns | Bāndŭsĭāe, | splēndĭdĭōr | vĭtrō
dūlcī | dĭgnĕ mĕrō | nōn sĭnĕ flōr|ĭbŭs
crās dō|nābĕrĭs haē|dō,
cuī frōns | tūrgĭdă cōr|nĭbŭs

(b) Sapphics

This is the favourite metre of Sappho. Catullus uses it in Poem 11, the last of the Lesbia poems (Part III, page 116) and in the first (Poem 51). In our selection of poems by Horace it is used in *Odes* 1, 20 (Part II, p.178).
The metre may be analysed as follows:
lines 1, 2, 3, extended base (– ⏑ – ⏓), choriamb (– ⏑ ⏑ –), clausula (⏑ – ⏓)
line 4 choriamb (– ⏑ ⏑ –), shortened clausula (⏓).

vīlĕ pōtā|bīs mŏdĭcīs | Săbīnŭm
cānthărīs, Gra͞e|că quŏd ĕgo īp|sĕ tēstā
cōndĭtūm lē|vī, dătŭs īn | thĕātrō
cūm tĭbĭ pla͞u|sŭs

(c) Alcaics

The favourite metre of both Alcaeus and Horace; in our selection *Odes* 2, 7 and 3, 26 – Part II, pp. 135 & 196.

ō sa͞epĕ mēcūm ∧ | tēmpŭs ĭn ūl|tĭmūm
dēdūctĕ Brūtō ∧ | mīlĭtĭa͞e | dŭcĕ,
quīs tē rĕdōnāvīt Qŭirītĕm
dīs pătrĭīs Ĭtălō|quĕ cāelō

⏓ – ⏑ – – ∧ | – ⏑ ⏑ – | ⏑ ⏓
⏓ – ⏑ – – ∧ | – ⏑ ⏑ – | ⏑ ⏓
⏓ – ⏑ – – – ⏑ ⏓
– ⏑ ⏑ – ⏑ ⏑ – | ⏑ – ⏓

In the first two lines there is always a caesura after the fifth syllable; the second half consists of choriamb + clausula.
The third line starts like the first two but there is no caesura after the fifth syllable and no choriamb in the second half; it is a very heavy line.
The last line moves fast; it might be analysed as a hemiepes (– ⏑ ⏑ – ⏑ ⏑ –) followed by a clausula (⏑ – ⏓).

(d) Hendecasyllables (i.e. lines of eleven syllables)

This is Catullus's favourite metre, used in the majority of his short poems; in our selection Poems 2, 3, 5, 9, 10, 13, 14, 46, 49, 50, 53.

It consists of Aeolic base (⏓ ⏓), choriamb (– ⏑ ⏑ –), extended clausula (⏑ – ⏑ – ⏓)

pāssēr | dēlĭcĭa͞e | mĕa͞e pŭēlla͞e
quīcūm | lūdĕrĕ, quem īn | sĭnū tĕnērĕ,
cuī prī|mūm dĭgĭtūm | dăre āppĕtēntī
ĕt āc|rīs sŏlĕt īn|cĭtārĕ mōrsūs

NB After the principles of scansion have been mastered, it is important to emphasize that Latin verse must be read with attention not only to scansion but also to the natural pronunciation of the Latin words. (The rules for the placing of the stress accent in the pronunciation of Latin are the same as in English.) Thus there are two rhythms being sounded at once, and the reader has to try to acknowledge the existence of both. This is not as difficult as it may seem. The most famous iambic pentatmeter in English has a trochee (– ⏑) in its fourth foot and a feminine ending, and yet does not lose contact with its basic iambic pulse:

Tŏ bē | ŏr nōt | tŏ bē: | thāt ĭs | thĕ quēst|ĭon:

In Virgil's hexameters, the clash between the stress of the scansion and the natural word stresses (marked ´) is usually resolved towards the end of the line:

lū́nă prĕ́|mīt sūā́|dēntqŭe că|dḗntĭă | sī́dĕră | sṓmnōs.

The above is a very soothing line. A far more agitated impression can be given by the conflict between the two stresses:

sēd cắdăt | āntĕ dĭ́|ēm || mĕ́dĭa | ī́nhū|mā́tŭs ăr|ḗnā.

If the final word of a hexameter is a monosyllable (apart from the verb **sum**), it creates an effect of dislocation or excitement:

Māssȳ́|līquĕ rū́|ūnt ĕ́quĭ|tēs ĕt ŏ|dṓră cā́|nūm vī́s.